# Catchment and River Basin Management

The central focus of this volume is a critical comparative analysis of the key drivers for water resource management and the provision of clean water – governance systems and institutional and legal arrangements. The authors present a systematic analysis of case study river systems drawn from Australia, Denmark, Germany, the Netherlands, the UK and the USA to provide an integrated global assessment of the scale and key features of catchment management.

A key premise explored is that despite the diversity of jurisdictions and catchments there are commonalities to a successful approach. The authors show that environmental and public health water quality criteria must be integrated with the economic and social goals of those affected, necessitating a 'twin-track' and holistic (cross-sector and discipline) approach of stakeholder engagement and sound scientific research.

A final synthesis presents a set of principles for adaptive catchment management. These principles demonstrate how to integrate the best scientific and technical knowledge with policy, governance and legal provisions. It is shown how decision-making and implementation at the appropriate geographic and governmental scales can resolve conflicts and share best sustainable practices.

**Laurence Smith** is Professor of Environmental Policy and Development in the Centre for Development, Environment and Policy, SOAS, University of London, UK.

**Keith Porter** is Adjunct Professor at Cornell Law School and the former Director of the New York State Water Resources Institute, Cornell University, Ithaca, USA.

**Kevin Hiscock** is Professor of Environmental Sciences, University of East Anglia, UK.

**Mary Jane Porter** is recently retired from the New York State Water Resources Institute, Cornell University, Ithaca, USA.

**David Benson** is Lecturer in Politics at the University of Exeter, UK, based at the Environment and Sustainability Institute (ESI) in Penryn, Cornwall.

# Earthscan Studies in Water Resource Management

For more information and to view forthcoming titles in this series, please visit the Routledge website: http://www.routledge.com/books/series/ECWRM/

# Catchment and River Basin Management

Integrating Science and Governance

**Edited by Laurence Smith,
Keith Porter, Kevin Hiscock,
Mary Jane Porter and David Benson**

LONDON AND NEW YORK

First published 2015 by Routledge

2 Park Square, Milton Park, Abingdon, Oxon OX14 4RN
711 Third Avenue, New York, NY 10017, USA

*Routledge is an imprint of the Taylor & Francis Group, an informa business*

First issued in paperback 2017

*British Library Cataloguing in Publication Data*
A catalogue record for this book is available from the British Library

*Library of Congress Cataloging in Publication Data*
Catchment and river basin management : integrating science and governance / edited by Laurence Smith, Keith Porter, Kevin Hiscock, Mary Jane Porter and David Benson.
pages cm. -- (Earthscan studies in water resource management)
Includes bibliographical references and index.
ISBN 978-1-84971-304-7 (hardback) -- ISBN 978-0-203-12915-9 (ebk)
1. Watershed management. 2. Water-supply--Management. 3. Watershed management--Case studies. 4. Water-supply--Management--Case studies.
I. Smith, Laurence (Laurence E. D.) editor of compilation.
TC413.C37 2015
333.73--dc23
2014042044

ISBN: 978-1-84971-304-7 (hbk)
ISBN: 978-1-138-30454-3 (pbk)

Typeset in Bembo
by Saxon Graphics Ltd, Derby

# Contents

# Figures

# Tables

# Boxes

# Contributors

**David Benson** is a Lecturer in Politics at the University of Exeter. His research, based at the Environment and Sustainability Institute (ESI) in Penryn, encompasses multiple areas at the interface between political and environmental sciences, most notably EU environmental and energy policy, comparative environmental politics and governance, and public participation in environmental decision-making.

**James Curatolo** resides in the Finger Lakes Region of New York State. He is Chairman of the Wetland Trust and Wetland Team Leader for the Upper Susquehanna Coalition.

**Kevin Hiscock** is Professor of Environmental Sciences at the University of East Anglia, Norwich and has 30 years of experience researching the impacts of land use and climate change on surface water and groundwater quantity and quality. His ongoing research is testing measures to mitigate diffuse pollution runoff from arable agriculture.

**Mark Horton** is the manager of Ballinderry Rivers Trust and its environmental consultancy business River Care Ltd. Previous to this, between 2004 and 2012 he coordinated projects for the Trust, including the RIPPLE project.

**Alex Inman** is an independent social researcher working in the field of integrated catchment management, both in the UK and internationally. Alex worked as a researcher and consultant in the RELU-funded project on Catchment Management for the Protection of Water Resources that led to this book.

**Tobias Krueger** is a junior professor at the Integrated Research Institute on Transformations of Human-Environment Systems of Humboldt-Universität zu Berlin. One strand of his research is the model-based support of land-water management with emphasis on uncertainty estimation and stakeholder participation. Tobias also advances methods development in these two areas.

**Mike Lovegreen** worked as manager of the Bradford County Conservation District in Pennsylvania for over 30 years, and now splits his time between project coordination for the district and work as Stream Team Coordinator for the Upper Susquehanna Coalition. He is also Leadership Development Program Coordinator for Pennsylvania State Conservation Commission, setting up programmes for conservation districts in the state.

**Stephen C. Maberly** is Head of the Lake Ecosystems Group at the NERC Centre for Ecology and Hydrology based at Lancaster. He has a broad expertise in lake ecology and a specific interest in carbon; its fixation by phytoplankton and macrophytes and its role in global cycles.

**Lisa Norton** is Head of the Land Use Group at the NERC Centre for Ecology and Hydrology (CEH) based at Lancaster. She has worked with CEH for the last 14 years and specialises in agricultural impacts on the environment at landscape scales. She manages the Countryside Survey, a national integrated monitoring scheme for the UK.

**Keith Porter** is Adjunct Professor of Law at Cornell University Law School. He was the US Principal in the international catchment management research project that led to this book. As Director of the New York State Water Resources Institute at Cornell from 1986 to 2007 he led inter-disciplinary research and outreach programmes in New York State, including the nationally significant New York City Watershed Protection Program.

**Mary Jane Porter** recently retired after long experience as an extension and water specialist with New York State Water Resources Institute (WRI) at Cornell University. Her work included management of the WRI Competitive Grants Program and co-management with the NYS Department of Environmental Conservation of the Hudson River Estuary Program and the NY Project WET (Water Education for Teachers).

**Laurence Smith** is Professor of Environmental Policy and Development in the Centre for Development, Environment and Policy at SOAS, University of London. His interests span natural resources, rural development and water resources management. He was Principal Investigator for the research project on Catchment Management for the Protection of Water Resources that led to this book.

**Diane Tarte** is Director of Marine Ecosystem Policy Advisors in Brisbane, Australia, and specialises in providing advice on policy and programmes addressing research and management of marine, coastal and catchment areas, with particular focus on ecosystem-based management of catchments, waterways and fisheries. From 2002 to 2010 she was Project Director for SEQ Healthy Waterways Partnership, leading a regional programme delivering science, monitoring, capacity-building and communications activities.

**Judith Tsouvalis** is a research associate working for the Leverhulme Trust on the 'Making Science Public' programme, based at the Institute of Science and Society (ISS), School of Sociology and Social Policy, University of Nottingham. As a human geographer she is interested in people's relations to nature and in the politics of nature, both of which she has empirically and theoretically explored in the contexts of forestry, farming and catchment management.

**Claire Waterton** is Senior Lecturer and Director of the Centre for the Study of Environmental Change (CSEC) within the Sociology Department at Lancaster University. Her research and teaching uses the theoretical approaches developed in Science and Technology Studies (STS) to explore contemporary environmental and policy problems.

**Nigel Watson** is a member of the Lancaster Environment Centre at Lancaster University. His research and teaching interests are in the area of water and environmental governance, and he advises governments and catchment groups on institutional arrangements and collaborative decision-making.

**Ian J. Winfield** is a member of the Lake Ecosystems Group at the NERC Centre for Ecology and Hydrology based at Lancaster. His research interests focus on the ecology of lake fish populations and communities and their assessment and management, with specific interests in rare fish conservation and sustainable fisheries.

# Foreword

Water is one of the most basic human needs. Protection of water supplies and the governance this entails must concern us all, whether that is at a global or local level. The complexity of this kind of challenge requires a holistic approach, drawing in not only researchers from across different disciplines but from a wide range of organisations and individuals. The Rural Economy and Land Use Programme (RELU) was underpinned by just such a philosophy. When it was launched in 2003, as an unprecedented collaboration between three research councils, it aimed to investigate the multiple challenges facing rural areas. One of these was the management of land and water use for sustainable water catchments. Other topics encompassed restoring public trust in food chains, tackling animal and plant disease, enabling sustainable farming in a globalised market, promoting robust rural economies and developing land management techniques to deal with climate change. None of this could be achieved without secure, sustainable water supplies for the benefit of people and of our environment. In this book, researchers who contributed to the RELU Programme and colleagues from across the world examine these complex issues and put forward innovative approaches that could help to address them.

*Professor Philip Lowe,*
*Director, Rural Economy and Land Use Programme*

# Acknowledgements

Research informing the conception and content of this book was undertaken with a Capacity Building Award and subsequent Research Project Award from the Rural Economy and Land Use Programme (RELU), a collaboration between the United Kingdom's Economic and Social Research Council (ESRC), the Biotechnology and Biological Sciences Research Council (BBSRC) and the Natural Environment Research Council (NERC). Additional funding of the RELU Programme was provided by the Scottish Government and the Department for Environment, Food and Rural Affairs. A second RELU Research Project Award supported establishment of the Loweswater Care Project (LCP) as described in Chapter 9.

The editors and authors very gratefully acknowledge the contributions made to the research and hence to this book by a host of catchment stakeholders and water sector professionals in the countries represented by the case study chapters and synthesis in this book. They are too numerous to mention by name, but particular thanks are owed to the farmers and other local residents who received our visits to their homes and businesses and fielded our many questions, and who participated in workshops and other discussion sessions.

Among professionals, in the USA particular thanks for their generous sharing of time, knowledge and experience are owed to Dean Frazier, Patricia Bishop, Steve Pacenka, Scott Cuppett, Katy Dunlap, Simon Gruber and Fran Dunwell. The time, support and information provided by co-authors in this book Jim Curatolo and Mike Lovegreen, and by their USC colleagues, were also immense. Valuable background research contributions were made by Sorell Negro, Julia Dobtsis and Michael Bowes. From Australia, Di Tarte and Eddie Hergerl provided invaluable information and advice, aided by colleagues from the Healthy Waterways Partnership. In Denmark, Gitte Ramhøj, Per Grønvald, Lise Kristensen, Lars Mortensen and Anne Jensen kindly shared their time and experience with us. In Germany, we were similarly aided and supported by Christina Aue, Onno Seitz and their colleagues; and in the Netherlands by Nico van der Moot, Auke Kooistra and colleagues. In Ireland, Mark Horton, Claire Cockerill, Ann Marie McStocker and colleagues facilitated study of the WWF RIPPLE project and work of the Ballinderry Rivers Trust. At Loweswater in England, the assistance provided by Ken Bell stands out among the participation and many contributions by local residents. Elsewhere in England we have learnt in particular from the work of the Westcountry Rivers Trust in south-west England, aided by Dylan Bright, Laurence Couldrick, Ross Cherrington and their colleagues; also from the work of the Broads Authority and Upper Thurne Working Group in East Anglia, aided by Andrea Kelly, Simon Hooton and colleagues. Other leading contributors to the RELU-funded research who provided invaluable assistance and insights include Hadrian Cook, Alex Inman, Jon Hillman and Andrew Jordan.

# Part I

# Overview

# 1 The challenge of protecting water resources

## An introduction and the purposes of this book

*Laurence Smith, Keith Porter, Kevin Hiscock, David Benson and Mary Jane Porter*

### The necessity of integrated catchment management

Healthy ecosystems, a clean environment and safe water supplies are vital to human health, quality of life and economic well-being, and thus to the overarching goal of sustainable development. It is a simple truism that our water resources are irreplaceable. They meet our needs for drinking, food and fibre production and hygiene, and they sustain the industries and ecosystems that support our livelihoods and lifestyles (Millennium Ecosystem Assessment, 2005). Protection of the natural ecosystems of river basins and the restoration of degraded water catchments are crucial to securing the world's water supplies, maintaining their quality, regulating floods, mitigating threats to water security from climate change, conserving biodiversity and enhancing cultural benefits and social values (Vörösmarty *et al.*, 2010; UNEP, 2011).

Beyond all other natural resources it is in our use of fresh water that our dependence on a healthy environment is most evident. Society has entered an era defined by the confrontation between natural limits to our resource use and growth in human demands in which the challenges of water resource management are at the forefront. These are global challenges as the amount of fresh water on Earth is finite. Achievement of social and economic development, improved social equity and stability, and the conservation of aquatic habitats and biodiversity all depend on our management of this scarce resource. Put simply, good water management is essential for sustainable development.

Despite this need, water resources continue to be degraded and although this is catalogued by an increasing frequency and volume of international meetings and reports, remedial actions at both catchment and river basin scale often remain inadequate. The following quotes highlight some of the challenges.

### *Water scarcity*

> More than 40% of the global population is projected to be living in areas of severe water stress through 2050.

There is clear evidence that groundwater supplies are diminishing, with an estimated 20% of the world's aquifers being over-exploited, some critically so.

(WWAP, 2014)

There are major uncertainties about the amount of water required to meet demand for food, energy and other human uses, and to sustain ecosystems. These uncertainties are compounded by the impact of climate change on available water resources.

(WWAP, 2012)

### Degraded water quality

Over 80 per cent of sewage in developing countries is discharged untreated directly into water bodies.

Industry is responsible for dumping an estimated 300–400 million tonnes of heavy metals, solvents, toxic sludge and other waste into waters each year.

Nutrient enrichment has become one of the most widespread water quality problems, severely degrading freshwater and coastal ecosystems.

The biodiversity of freshwater ecosystems has been degraded more than any other ecosystem, including tropical rainforests.

(UN Water, 2011)

### Nutrient pollution

The sustainability of our world depends fundamentally on nutrients. In order to feed 7 billion people, humans have more than doubled global land-based cycling of nitrogen (N) and phosphorus (P).

The world's N and P cycles are now out of balance, causing major environmental, health and economic problems that have received far too little attention.

Unless action is taken, increases in population and per capita consumption of energy and animal products will exacerbate nutrient losses, pollution levels and land degradation, further threatening the quality of our water, air and soils, affecting climate and biodiversity.

(Sutton *et al.*, 2013)

Failures in the sustainable management of water resources are evident in most regions of the world. Extreme and well-known examples include the Aral Sea, diminished in volume and area by abstraction from its tributaries, and the 'Dead Zone' of the Gulf of Mexico. The latter occurs regularly and consists of a swathe of shallow coastal waters up to 22,000 square kilometres in area (equivalent to the state of New Jersey) within which there is not enough oxygen in the water to support marine life. This is caused in large part by the run-off from farms in the mid-west of the USA, from where excess nitrogen and phosphorus lost from liberal use of fertilizers on farmland is carried by the Mississippi to the Gulf. Resultant blooms of algae deprive the water of oxygen as they die and decompose.

Of equal if not greater significance in aggregate to such 'flagship environmental disasters' are the less extreme but globally widespread examples of overabstraction of surface and groundwater, deterioration in inland and coastal water quality, degradation of aquatic and wetland ecosystems and increased incidence of damaging floods.[1] Of particular concern in this book is the deterioration in water quality caused by diffuse sources of pollution from human activities in rural areas. However, this cannot be considered in isolation from other aspects of water and land management in our river basins and catchments.

Rain, hail and snow are the ultimate source of water but once it starts to melt, infiltrate or run off, the way water moves, is stored, is lost or degraded depends mainly on the characteristics of the land and how it is managed. The sustainable development challenge at the 'heart' of this book thus concerns how best to protect and conserve water within the landscapes in which people live, work and play. In other words, it is about managing land and water resources to achieve the multiple aims of clean and safe water supplies, profitable production of food and other agricultural commodities, viable rural livelihoods, businesses and settlements, and natural spaces for recreation, physical and psychic health, and attractive lifestyles. This requires the sustained conservation of healthy and diverse ecosystems and the goods and services they provide. However, despite the passage of a quarter-century since the Brundtland Report,[2] neither the principles of sustainable development, nor the principles of the 'ecosystem approach'[3] have yet to be comprehensively adopted in catchment and river basin management.

Inherent in these challenges are many complex complementarities and trade-offs. These prompt a comprehensive and integrated approach, and raise questions about the scale and scope of necessary action. In terms of scale, the river basin is typically the natural geographical unit to consider. A river basin encompasses the area drained by a river and its tributaries, from the source of each stream to a final destination in the sea, an estuary or inland lake. Most of Earth's land surface falls within the area of river basins. In comparison the terms 'catchment' and 'watershed' are commonly used to refer to areas of land defined by the sub-basins of tributaries within a river basin. Many such smaller catchments or watersheds can exist within a basin. More precisely a watershed refers to the divide that separates one drainage area from another, although in some countries, including the United States and Canada, the term is also applied to the river basin or catchment area itself.

There are several reasons why it is considered logical to take a river basin or catchment as the spatial unit for analysis, planning and management of land and water resources. The river and its tributaries are common to all parts of the basin and water users within the basin are interdependent insofar as the actions of one person can affect the amount and quality of water available to others. In particular, upstream water users and land managers can affect the volume, quality or seasonality of water available to those downstream. For example, draining upland peat bogs or other wetlands can remove water storage capacity and natural mechanisms that modulate the flow of a river and assimilate waste. Similarly, deforestation and change in upstream land use may affect the frequency and severity of downstream floods, while also increasing soil erosion and the sediment and other pollutants carried by the river system. Such mechanisms are also manifest as catchment to coast linkages such that changes in upstream land and water use can impact on near-shore coastal ecosystems. Pollution largely conveyed by rivers from land-based sources can account for as much as 80 per cent of all marine pollution. Overall the challenges of water scarcity, flood risk and diminished water quality each impose the need to manage water at each stage from its source to the sea. In addition, if it is true that water users and land managers within a catchment can perceive their dependence on a common resource and understand their interdependence with others, then a basis for resolution of conflicts of interest and competition over resources may exist. A catchment or basin thus appears to provide both a natural unit for strategic planning and management, and a potential 'forum' for assessment of resources, trade-offs and the incidence of impacts. This introduces the duality of technical assessment and stakeholder deliberation that is a core theme of this book.

Although these arguments for planning, analysis and management at a catchment or basin scale are clear there can be exceptions and obstacles. Exceptions may include transboundary effects arising from groundwater flows or inter-basin water transfers. In terms of obstacles it is usual to find that watersheds and thus the boundaries of catchments do not match existing administrative and political boundaries. Such institutional boundaries were usually established in eras before the relative scarcity or poor quality of water resources became a leading concern. Also, while individuals can be expected to be responsive to issues and trade-offs within their immediate locality, it is not clear that they will always be either cognizant of, or responsive to, objectives set at a larger catchment or basin scale.

Despite the importance of catchments and river basins it can thus be difficult to determine the best scale at which to manage land and water resources, or at which to implement programmes and policies for their improvement. This will depend on a range of factors and a complexity which can be compounded by the timescale over which problems may materialize and the cumulative nature of impacts that can occur. As we will explore further in this book local management of land and water at a catchment or sub-catchment scale is the natural default situation and has many advantages, but the logic of water resource management at a whole catchment or basin scale will also require interventions at this higher scale. In turn this will require appropriate and effective multi-agency and multi-level governance arrangements.

The logic of catchment management also reveals that any single management intervention in a catchment or basin can be expected to have foreseen consequences and side effects, both of which will be subject to uncertainty. Given such complexities and interdependencies the logic of assessment and management at a catchment scale extends to the need to be holistic and comprehensive in scope. The concept of integrated catchment management thus includes integration of the management of landscapes, waterways, lakes, estuaries and coastal ecosystems within a river basin. It encompasses the management of both water quantity and quality, and diverse uses of water by the public, industry, agriculture and other land uses, and the energy sector. It requires data and scientific understanding that is integrated both across disciplines and with the location-specific knowledge of local residents and resource users. As noted above integrated catchment management must be coordinated across the tiers and divisions of administration and civil society that exist. In terms of objectives, integrated catchment management requires that economic and social goals are aligned with ecological and environmental outcomes. In implementation it requires that voluntary efforts coincide in objectives and are coordinated in implementation with the programmes and projects of the private and public sectors.

Catchment management exhibiting some or all of these features of scale and scope has become a global phenomenon, but with different forms emerging in different contexts (Benson *et al.*, 2013) – for example, 'watershed management' in the USA prompted by the provisions of the federal Clean Water and Safe Drinking Water Acts; 'integrated catchment management' in Australia under the Landcare programme and latterly the National Heritage Trust and Caring for our Country initiative; 'river basin management' as required by the Water Framework Directive in the European Union; and the Integrated Water Resources Management (IWRM) paradigm adopted by the United Nations and other international bodies. Each of these can be contrasted to prior governance regimes for water that can typically be characterized as technocratic and engineering led, primarily public sector, and focused on single-sector objectives of water abstraction and supply, hydropower, navigation or flood control (Molle, 2009).

This global and 'paradigmatic' change in management philosophy and practices potentially provides multiple and diverse cases on which to draw for learning and lesson transfer between locations and jurisdictions. A selection of leading cases provides the basis for this book and its attempt to draw lessons for catchment management, and specifically the control of diffuse pollution in rural areas.

## The objectives, focus and scope of this book

In this book 'catchment management' is used as a generic term to refer to the management of water and to the relevant management of land uses and built infrastructure at a catchment and sub–catchment scale. As such use of the term overlaps with a range of related concepts including river basin management planning, watershed planning, integrated catchment management and relevant aspects of IWRM.

The subject of this book is how best to protect, conserve and manage water resources at their source and at the scale of a catchment or watershed. Issues of scale are very important and raise challenges for the transferability of recommendations and best practice. This book focuses on small- to medium-scale catchments for surface and groundwater. Challenges at a whole river basin scale for major rivers, including international transboundary issues, are not directly addressed although some of the lessons drawn here are potentially transferable to such cases. The issues and approaches that are considered at a smaller scale are certainly relevant as 'building blocks' and potential delivery mechanisms for the challenges of large basins.

This book seeks to provide an innovative perspective by adopting an inter-disciplinary approach to critical comparative analysis of drivers and modalities for catchment-based water resource management that include governance systems and institutional and legal arrangements. Catchment management crucially depends on these drivers and modalities. Case studies in the book allow a comparison of catchment scientific approaches, governance systems and institutional arrangements for the management of water resources, drawn from case examples in the USA, Australia and north-west Europe. This collective experience and understanding is supported by lessons culled from the wider international literature to attempt an integrated and globally applicable synthesis of the scope and commonalities of catchment management.

Focusing on water as a natural resource, the primary purposes for catchment management are the efficient allocation and use of a catchment's water resources and cost-effective application of measures to protect catchment ecosystems and the quantity and quality of the water that these produce. A further breakdown of water management objectives according to water use by sectors of the economy could include: water supplies for people and animals; water for agricultural (including aquaculture) and industrial production; water for nature conservation and capture fisheries; navigation; recreation; power production; flood control; fire protection; and waste disposal.

Overabstraction, flood risk and water quality are almost universal concerns. Water pollution comprises point and diffuse (or non–point) sources of contamination including discharges from wastewater treatment and industry, surface run-off from fields, seepage of nutrients and other contaminants from soil into groundwater, stream bank erosion and discharges from dispersed and numerous minor point sources such as field, farmyard and urban drains. Diffuse or non–point source water contamination is the most difficult to overcome and its control is a primary concern of this book. A point source can be defined as a discrete and discernible conveyance of wastewater such as pipes, ditches, channels and other means of conveying water. Point source water pollution is potentially amenable to solutions based on the pre-discharge treatment of wastewater, and implementation of such solutions requires a combination of regulation, technology and investment. This is feasible where an economy is sufficiently developed, but challenges may remain in terms of securing the necessary political will, prioritization and financing.

In contrast diffuse or non-point source water pollution is the release of pollutants from dispersed activities and sources across the landscape. It comprises true non-point source contamination and that which arises from many individual and relatively minor point sources. Examples of true non-point sources are seepage of nutrients from soil into groundwater and sheet run-off from the surface of fields or via numerous sub-surface field drains. Examples of multiple minor point sources are field ditches and farmyard drains or surface water drains from houses, other buildings and hard standings. The distinction made here illustrates that the classification of a pollution source as point or non-point (diffuse) may often not be a clear one. For practical purposes, particularly in rural areas, it may be better to recognize a spectrum of pollution sources from the truly diffuse to the large-scale and concentrated discharges of industrial units, wastewater treatment plants or intensive livestock farming operations. Although more diffuse sources may individually have negligible effect on the water environment, at the scale of a catchment or water body they may have a significant impact in aggregate.

Before industrialization began in the early nineteenth century the main causes of water pollution were untreated human waste and organic and inorganic pollution from artisanal industry. Since then rapid technological development, economic development and population growth, and industrial and urban expansion have intensified and broadened pollution problems from agricultural and other rural sources, road and urban stormwater run-off, sewage and industrial effluents. The challenges also continue to evolve, and for example, new chemicals such as pharmaceuticals which may not be amenable to standard treatments of wastewater discharge are increasingly recognized as a problem that is increasing in its severity.

A dominating sub-set of diffuse water pollution issues are those associated with eutrophication; a term used 'to describe the complex sequence of changes in aquatic ecosystems caused by an increased rate of supply of plant nutrients to water' (Schindler and Vallentyne, 2008). 'Cultural eutrophication' (Schindler and Vallentyne, 2008) can be rapid and is the enrichment of water with nutrients, principally compounds of phosphorus and nitrogen, from human activities. This contrasts with the very slow process of natural eutrophication of lakes and other water bodies as they accumulate sediment from their catchments. The main human sources of nutrients are inadequately treated sewage, arable and livestock farming, conversion of forested land to other uses, and other sources including garden fertilizers and pets (Schindler and Vallentyne, 2008).

Collectively such pollution causes environmental damage and carries health risks and other economic costs. Nutrient loads to water bodies from untreated human sewage and livestock manure are usually accompanied by significantly raised concentrations of coliform bacteria or protozoans such as *Cryptosporidium* or *Giardia*, carrying the risks and costs of waterborne disease. The nutrients themselves promote the growth of many species of alga, the populations of which can be used to define the trophic status of a lake or other water body. In the worst cases eutrophic lakes and slow-moving rivers can become dominated by Cyanobacteria, commonly known as blue-green algae. Such species are poor food for higher aquatic animals and can proliferate to form blooms or scums on the water surface. These are generally unsightly and unpleasant, while some species can release toxins harmful to animals and people (Schindler and Vallentyne, 2008). Such algal blooms can give an 'off-taste' and turbidity to drinking water, require cleaning from water intake filters and pipelines, and render bathing and recreational areas unpleasant or unusable. Problems caused by excessive nutrients are likely to be worsened during periods of lower water flow and higher temperatures in summer months. As noted above, oxygen depletion of both inland and coastal water can also result from algal blooms induced by nutrient pollution. Though photosynthesis by the algae may raise dissolved oxygen during the day, at night a dense algal bloom will reduce dissolved oxygen through respiration, and further oxygen depletion in the water column occurs when algal cells die, sink towards the bottom and are decomposed by bacteria. Ultimately a 'dead zone' arises when oxygen depletion progresses to hypoxia and death for all fish and invertebrates.

In recent years a combination of legislation, investment in wastewater treatment plants and de-industrialization has improved the quality of surface waters in some countries by controlling or eliminating discharges from point sources. This process is generally more advanced in developed economies, whereas developing countries are experiencing different stages and combinations of the problems identified above. In rapidly developing countries such as China, India, Mexico and Brazil, agricultural intensification and unprecedented rates of urbanization and industrialization have caused water pollution problems on a huge scale (Shiklomanov, 1997).

This book draws on experience from the north-eastern United States of America, south-eastern Queensland in Australia and north-west Europe. In each of these regions at least some reduction in pollution from point sources has been achieved through regulation, investment and technological improvements in the treatment of wastewaters (and as a result of de-industrialization in some locations), but water quality problems remain because of diffuse pollution derived from current and past land use (agricultural and urban) plus atmospheric deposition. These are global problems wherever farming is sufficiently intensive and relatively dense human populations are served by inadequate sewage treatment facilities. Data shows that in OECD countries the pollution of rivers with nitrogen generally increases with the usage of nitrogenous fertilizers per hectare of arable land, although there is considerable geographical dispersion around this trend (United Nations, 2010).

Diffuse water pollution poses particular challenges for public policy, the implementation of control strategies, best management practices, and scientific research and analysis. Innovative management approaches are required as solutions ultimately require behavioural change by many actors and thus a broad societal response. They must also be flexible and adaptive to stochastic catchment conditions and to long-term trends. They must be integrated with the management of abstraction and flows, and of flood risk. Internationally new models of governance for difficult land and water resource management problems of this nature are emerging that recognize and seek to incorporate such factors. This nexus of challenges and the means for their solution provides the focus for this book.

### The challenges and uncertainties of controlling diffuse water pollution

Given natural limits to resource use and growing human demands, solely technological and reductionist approaches to the management of water and other natural resources are being questioned in the face of environmental degradation, and even collapse of ecosystems (Allan *et al.*, 2008). Many natural systems are inherently dynamic and complex, and knowledge of how they work and what determines outcomes is incomplete. At the same time decisions on how people use and conserve natural resources are subject to diverse legitimate but often competing values, resulting in a wide range of objectives and interests needing to be met. These are the characteristics of a 'wicked problem' and diffuse water pollution provides a clear example of such a challenge (Smith and Porter, 2010). This is explained further below.

Consider first the scientific uncertainty that may exist for land and water users, and for environmental scientists and managers. The absorptive capacity of surface and groundwater for diffuse pollution is a 'common pool' resource and stopping the use of that resource for disposal of pollutants (whether deliberate or a consequence of lack of awareness and control measures), through measures such as imposing a zero-discharge policy for the non-point sources, is prohibitively costly if not impossible. Polluters make multiple and interdependent uses of land and water resources and are numerous, dispersed and often remote. Typically they are not fully aware of the environmental consequences of their actions, and they will vary in their perception of both water quality status and its value. For scientists and managers there is uncertainty about pollution sources, pathways and impacts, about the occurrence of spatially and temporally stochastic pollution events, and about the consequences of any given pollutant loading for ecosystem, economic and public health outcomes. Individual discharges are for most practical purposes unobservable and unverifiable and thus cannot serve as the basis of regulation. Even if a source is observable, the transmission path from source to receptor involves many unknown variables such as rainfall, soil type, microbial activity and level of groundwater table, preventing the ability to relate the effluent to both its source and impact. This, together with a time lag between emission of a diffuse pollutant and its appearance in a receptor stream, lake or aquifer exacerbates the difficulty of establishing a link between the source and the ambient level of pollution (O'Shea, 2002).

The efficacy of control measures and the most cost-effective approaches for monitoring are also uncertain. As noted, solutions to diffuse water pollution ultimately require behavioural changes on the part of land users, residents and other actors in rural and urban areas which are inherently difficult to achieve and sustain. Catchments are also heterogeneous and the data available to diagnose problems and to design and target prevention or control strategies are rarely adequate. These uncertainties are not all equal relative to the management purposes at issue. Determining the priority of uncertainties and the relative weight they merit is a key aspect of the development of solutions (Smith and Porter, 2010).

Design and implementation of pollution prevention and mitigation measures may require both generic and location specific bio-physical research, together with socio-economic assessments of their potential impacts. Examples include investment in community wastewater treatment systems, improvements to the sewage systems of remote rural homes, management practices on farms to contain animal manures and optimize plant and animal nutrient regimes, and changes to land use and landscape. The outcomes of all such technical solutions will depend on stochastic catchment and climatic conditions and longer-term trends in influences such as market opportunities for farm enterprises and other economic activity, or climate change.

Complexity and uncertainty also arise because pollution prevention and mitigation measures will usually themselves have other indirect benefits or costs. This emphasizes the need for a holistic, comprehensive and adaptively integrated approach to catchment management. Changes in land use and management practice, for example, may also provide improvements in habitat for non-aquatic biodiversity, contribute to downstream flood alleviation or alter the existing balance of greenhouse gas absorption or emission.

Another source of technical complexity for managers is manifest in the form of institutional issues. Spatial patterns of human and ecological water use and waste disposal, and the bio-physical boundaries of catchments and ecosystems, rarely coincide with administrative and other legal jurisdictions. A political and administrative fragmentation typically exists that can be a hindrance to coordinated and catchment-scale water management and protection measures. Legal authorities for water supply and protection may span multiple agencies and levels of government, prompting questions of whether and how the existing governmental and institutional framework can adjust to accommodate catchment-level disparities of geography and jurisdictions – for example, whether transboundary collaboration, and even new institutional structures, may be needed at scales ranging from local government to international river basin authorities. The technical problems of catchment management are also cross-sectoral, spanning the responsibilities of agencies for agriculture, forestry, fisheries, highways, planning, waste disposal, building regulation, flood control and the environment. This reinforces a need for inter-agency communication and coordination (Smith and Porter, 2010), and for the capability for cross-sectoral and multi-disciplinary assessment and planning.

Next consider uncertainty about society's values and objectives. Most people can be expected to express a desire for higher water quality and the other benefits that may stem from integrated and effective catchment management, but there are inherent trade-offs that may moderate this preference in varying degree for different groups. As a consequence the willingness of different groups to change behaviour or to bear some or all of the costs of pollution mitigation will vary. Diffuse pollution is an externality of land uses that produce goods, homes, livelihoods and landscape attributes that sustain rural communities, are generally desired by society, and in some incidences can be recognized as 'public goods'. It is thus valid to ask whether reducing water pollution will compromise the viability of the rural economy, and key questions that follow are: who should pay the costs of water protection, and who benefits most from pollution reduction and enhanced water quality? For example, should the burden of preventing water pollution fall on rural land users upstream for the benefit of a downstream urban area? Or should farmers in a water supply catchment area be required to achieve a higher standard of care in their management practices than that required of farmers in other catchments? Such questions raise issues of equity and social justice as well as economics. The answers are not immediately self-evident and will be contested. Assessment and planning to control diffuse water pollution must therefore extend beyond farms and wastewater treatment plants to consider possible trade-offs relating to all aspects of land and water use, landscape heritage and the rural economy, including the viability of settlements, transport networks, habitats and recreational activities (Cook and Smith, 2005).

From a policy perspective diffuse water pollution can also be understood as a case of 'market failure'. The absence of market incentives to reduce pollution motivates intervention by governmental and non-governmental entities to modify the incentives signalled by markets that determine behaviour. As explained above, unlike point source pollution, the temporal and spatial nature of diffuse pollution renders its complete monitoring and regulation impractical. Even if the sources of diffuse pollution can be definitively identified, the monitoring and enforcement costs of an approach based on regulation alone are likely to be prohibitive. As a consequence policymakers should be prepared to utilize a combination of measures that includes: regulation; economic incentives and voluntary agreements with land users; self-regulation based on enhanced knowledge and cultural changes; advisory and education campaigns; and direct land management strategies that may require land acquisition and change or restriction on use. This policy instrument 'mix' further adds to the complexity identified above, and the central challenge can be recognized as one of how to determine and implement the best combination of measures for a specific catchment, given local conditions and preferences, and wider national or transnational priorities and policy constraints (Cook and Smith, 2005). The most appropriate method of pollution control will depend upon, among other things: the information available; the type of resource to be regulated; the degree of uncertainty that exists; the costs of damage; the number of polluters to be controlled; and monitoring costs and transactions costs (O'Shea, 2002). Each case must be decided on its own merit, and scientific capability and an adequate technical knowledge base must be developed alongside the necessary governance arrangements for condition and threat assessment, planning, appraisal of proposed measures and implementation.

## The structure of this book

The narrative of this book falls into three main sections. Chapters 1 and 2 provide an introduction and overview. This first chapter has set the scene by defining and scoping some of the key challenges and characteristics of catchment management. Chapter 2 extends and deepens this analysis, providing a framework in the form of key questions for use in analysing case-study experience and establishing premises for the required functions and institutional arrangements of successful catchment management programmes.

The second section of the book provides description and analyses of nine catchment management case studies. These have been selected for their landmark innovation and achievements to date. Each chapter documents a case study (three cases are considered in Chapter 7), and a key premise explored is that despite the diversity of jurisdictions and catchments there are observable commonalities to a successful approach. Consistent comparative analysis of diverse case studies outcomes is challenging but this is attempted through use of the key questions identified in Chapter 2 and a common narrative framework for description of each case.

In Chapters 3 to 5, three leading watershed programmes in the USA are described and analysed. These are the Upper Susquehanna River Basin, the New York City watershed and the Lower Hudson River Valley. Chapter 6 turns to south-east Queensland in Australia and describes the evolution and outcomes of its Healthy Waterways Partnership. Chapter 7 turns to three cases from north-west Europe for which protection of groundwater has been the primary concern. Chapter 8 describes the development and implementation of a community-led catchment planning process in Northern Ireland, and Chapter 9 tells the story of how community-based learning was central to the development of the Loweswater Care Programme in north-west England.

The third and final section of the book attempts to draw out the commonalities of principles, process and governance that have contributed to success in the case studies. Integrating the relevant scientific, economic and social perspectives that emerge from the case studies this section identifies drivers of change, essential actions and approaches, and the key aspects of governance that include policy, legislation and other institutional and organizational arrangements. It is hoped that this final synthesis can support decision-making and implementation at the appropriate geographic and governmental scales to resolve conflicts and share best sustainable practices in catchment management.

Our aim is that the style and content of this book should be interesting and informative for catchment management practitioners and all catchment stakeholders. Thus use of citations and reference lists has been limited to important sources of information and evidence, but some reference is also made, particularly in Chapter 2 and in the concluding chapters, to theory and concepts from a wider academic literature that will provide a resource for further reading for those interested in pursuing this.

## Notes

1　The biennial report 'The World's Water' (eight editions to date, 1999 to 2014) by the Pacific Institute, and the United Nations 'World Water Development Report' series (five editions to date, 2003 to 2014) provide examples of water management challenges globally and data on water resource use and management.

2　In 1987 the Brundtland Report defined sustainable development 'as development that meets the needs of the present without compromising the ability of future generations to meet their own needs'. Then Agenda 21 was adopted by more than 178 governments at the United Nations Conference on Environment and Development held in Rio de Janeiro in 1992 as a comprehensive plan of action to be taken globally, nationally and locally in every area in which there are human impacts on the environment. Today the principle of sustainability is taken to imply use of resources at rates that do not exceed the capacity of our planet to replace them. For example, water consumed at a rate that can be replenished by basin inflows and rainfall, greenhouse gas emissions balanced by carbon fixation and storage, a halt to soil degradation and biodiversity loss, and no accumulation of pollutants in the environment; all achieved while economic activity maintains viable and healthy livelihoods and opportunities for future generations. Sustainability also entails resilience, for example, such that aquatic ecosystems and the services they provide are robust to transitory shocks and stresses (Government Office for Science, 2011).

3　The 'ecosystem approach' also gained international recognition at the 'Earth Summit' in Rio de Janeiro in 1992 when it was adopted as a foundation of the Convention on Biological Diversity (United Nations, 1992). It became integral to international environmental commitments made at the 2002 World Summit on Sustainable Development in Johannesburg (United Nations, 2002), and formed the conceptual basis for the Millennium Ecosystem Assessment conducted from 2001 to 2005 (Millennium Ecosystem Assessment, 2005).

## References

Allan, C., Curtis, A., Stankey, G. and Shindler, B. (2008) 'Adaptive management and watersheds: A social science perspective', *Journal of the American Water Resources Association*, vol. 44, no. 1, pp. 166–174.

Benson, D., Jordan, A. and Smith, L. (2013) 'Is environmental management really more collaborative? A comparative analysis of putative "paradigm shifts" in Europe, Australia, and the United States', *Environment and Planning A*, vol. 45, no. 7, pp. 1695–1712.

Cook, H. and Smith, L. E. D. (2005) 'Catchment management – Relevant in developed and developing countries', *Waterlines,* vol. 24, no. 1, pp. 2–3.

Government Office for Science (2011) *Foresight. The Future of Food and Farming*, Government Office for Science, London.

Millennium Ecosystem Assessment (2005) *Ecosystems and Human Well-Being: Wetlands and Water Synthesis,* World Resources Institute, Washington, DC.

Molle, F. (2009) 'River basin planning and management: The social life of a concept', *Geoforum,* vol. 40, no. 3, pp. 484–494.

O'Shea, L. (2002) 'An economic approach to reducing water pollution: Point and diffuse sources', *The Science of the Total Environment,* vols. 282–283, pp. 49–63.

Schindler, D. W. and Vallentyne, J. R. (2008) *The Algal Bowl,* University of Alberta Press, Edmonton.

Shiklomanov, I. A. (1997) *Comprehensive Assessment of the Freshwater Resources of the World: Assessment of Water Resources and Water Availability in the World,* World Meteorological Organization and Stockholm Environment Institute, Stockholm.

Smith, L. E. D. and Porter, K. S. (2010) 'Management of catchments for the protection of water resources: Drawing on the New York City watershed experience', *Regional Environmental Change,* vol. 10, no. 4, pp. 311–326.

Sutton, M. A., Bleeker, A., Howard, C. M., Bekunda, M., Grizzetti, B., de Vries, W., van Grinsven, H. J. M., Abrol. Y. P., Adhya, T. K., Billen, G., Davidson, E. A, Datta, A., Diaz, R., Erisman, J. W., Liu, X. J., Oenema, O., Palm, C., Raghuram, N., Reis, S., Scholz, R. W., Sims, T., Westhoek, H. and Zhang, F. S., with contributions from Ayyappan, S., Bouwman, A. F., Bustamante, M., Fowler, D., Galloway, J. N., Gavito, M. E., Garnier, J., Greenwood, S., Hellums, D. T., Holland, M., Hoysall, C., Jaramillo, V. J., Klimont, Z., Ometto, J. P., Pathak, H., Plocq Fichelet, V., Powlson, D., Ramakrishna, K., Roy, A., Sanders, K., Sharma, C., Singh, B., Singh, U., Yan, X. Y. and Zhang, Y. (2013) *Our Nutrient World: The Challenge to Produce More Food and Energy with Less Pollution,* Global Overview of Nutrient Management, Centre for Ecology and Hydrology, Edinburgh on behalf of the Global Partnership on Nutrient Management and the International Nitrogen Initiative.

UN Water (2011) *Water Quality: Policy Brief,* United Nations, New York.

UNEP (2011) *Towards a Green Economy: Pathways to Sustainable Development and Poverty Eradication,* United Nations Environment Programme, New York.

United Nations (1992) *Convention on Biological Diversity,* United Nations, New York.

United Nations (2002) *Report of the World Summit on Sustainable Development,* Johannesburg, South Africa, 26 August–4 September 2002, United Nations, New York.

United Nations (2010) *Trends in Sustainable Development: Towards Sustainable Consumption and Production,* Department of Economic and Social Affairs, Division for Sustainable Development, United Nations, New York.

Vörösmarty, C. J., McIntyre, P. B., Gessner, M. O., Dudgeon, D., Prusevich, A., Green, P., Glidden, S., Bunn, S. E., Sullivan, C. A., Reidy Liermann, C. and Davies, P. M. (2010) 'Global threats to human water security and river biodiversity', *Nature,* vol. 467, no. 7315, pp. 556–561.

WWAP (2012) *The United Nations World Water Development Report 4: Managing Water under Uncertainty and Risk,* United Nations World Water Assessment Programme, UNESCO, Paris.

WWAP (2014) *The United Nations World Water Development Report 2014: Water and Energy,* United Nations World Water Assessment Programme, UNESCO, Paris.

# 2 Key questions about catchment management

*Laurence Smith, David Benson and Keith Porter*

## Introduction

This chapter extends and deepens the identification of catchment management challenges introduced in Chapter 1 and provides a framework for comparison and analysis of the nine case-study examples of catchment management programmes in Chapters 3 to 9. Five main questions are used to structure this chapter and the description of each of the case studies. The issues addressed by the questions are interdependent and, as well-illustrated by the case studies, often evolve iteratively over time. They are discussed further in the sections that follow the list of questions below, along with important sub-questions and issues.

1  **What are the context and drivers for catchment management?**
   (Why did it start? What factors influence when and how a catchment management programme begins? What factors promote an integrated approach?)

2  **How can a catchment management programme get started?**
   (How did it start? What factors influence how catchment management is initiated and evolves?)

3  **What approaches and tools are needed for catchment management?**
   (Which approaches and methods work? How can science be integrated into effective catchment management?)

4  **How do things get done?**
   (How can successful catchment management be kept going? What governance arrangements sustain effective catchment management?)

5  **What are the outcomes of catchment management?**
   (How do we know that it works? What are appropriate measures of effectiveness and success? What capabilities must be demonstrated and other criteria met for a catchment management programme to be deemed successful and sustainable?)

## What are the context and drivers for catchment management?

An important starting point in any investigation is to understand the drivers that can initiate and sustain catchment management initiatives. These drivers can arise from the regulatory environment in a manner that can be characterized as 'top-down'. For example, legislation and standards may be set for the management of water quantity and quality to protect and conserve supplies for public use, protect public health, achieve environmental conservation or restoration, or to mitigate flood risk. The form that a catchment management programme takes will be a response to such regulation and is likely to be shaped by the goals of the legislation and by the prevailing governance arrangements and capacities. Such 'top-down' drivers may or may not promote the integration across issues and sectors generally recognized as an essential principle for integrated catchment management.

In contrast, catchment management can also arise from community-based initiatives in a manner that can be characterized as 'bottom-up'. Local concerns, priorities and 'triggers' for action can similarly encompass concerns about ability to abstract water, public health, environmental degradation and flooding. The form that catchment management takes in this case will first be shaped by the initial focus, local capacities and organizational and institutional arrangements at local level. This form will then increasingly interact with, and be further shaped by, higher-level institutions and governance arrangements as it evolves. If successful a catchment management initiative may also grow in scope, scale and importance. If genuinely driven by local needs and priorities such 'bottom-up' catchment management may be more likely to integrate the community's economic and social goals with water management objectives, and perhaps to integrate across water issues and sectors, at least at local level. This book provides the opportunity to assess the nature and influence of drivers for catchment management in a range of case studies and from this to draw lessons for the initiation of programmes elsewhere.

### What are the drivers for integrated catchment management?

As noted above an important sub-question is whether the 'top-down' or 'bottom-up' origins of a catchment management programme promote an integrated approach. River basins and catchments are dynamic and complex systems where water flows from land in the upper parts of the catchment, through lakes, rivers and groundwater towards estuaries and the sea. There are many interdependent natural processes in a drainage basin, and any human activity can affect the status of water and its ecology within the basin. Complexity also arises from the need to encompass hydrological cycles, biological systems and multiple human needs and values derived from use of land and water. Human activities are also interdependent and actions focused on a single water use, interest group or sector may affect others. As noted in Chapter 1 the logic of planning and managing water resources at a river basin or catchment scale is consequently well-established and widely accepted. However, the environmental, economic and social implications of change in a catchment can also extend beyond its boundaries. It follows that catchment management may often need to go beyond the geography and natural processes of a catchment and also span boundaries of self-interest, sector, discipline and governance.

It follows that catchment management should be considered holistic, comprehensive and inclusive by definition. Recognition of this is widely established, for example:

> river basin management (RBM) consists of all activities aiming at a better functioning of the river basin, including the water system and the land in the basin in as far as it is affected by or has an impact on the water system.
>
> (Mostert, 1999, p. 564)

> Integrated river basin management (IRBM) is the process of coordinating conservation, management and development of water, land and related resources across sectors within a given river basin, in order to maximise the economic and social benefits derived from water resources in an equitable manner while preserving and, where necessary, restoring freshwater ecosystems.
>
> (WWF, 2013)

> Integrated catchment management (ICM) is a process through which people can develop a vision, agree on shared values and behaviours, make informed decisions and act together to manage the natural resources of their catchment. Their decisions on the use of land, water and other environmental resources are made by considering the effect of that use on all those resources and on all people within the catchment.
>
> (Murray–Darling Basin Ministerial Council, 2001, p. 1)

> Integrated water resources management (IWRM)[1] is a process which promotes the coordinated development and management of water, land and related resources in order to maximize the resultant economic and social welfare in an equitable manner without compromising the sustainability of vital eco-systems.
>
> (GWP, 2000, p. 22)

> Integrated Land and Water Resources Management (ILWRM) and 'Resilience-based ILWRM' are more recent and systems-oriented approaches that broaden the scope of IWRM to include social–ecological interactions and the uncertainties of global change.
>
> (Rockström *et al.*, 2014)

Common criticisms of such calls to be holistic and integrated are that a wide gap remains between theory and practice, application is typically inadequate and unfinished, and empirical evidence of success is limited or lacking. IWRM in particular has been said to remain a normative paradigm without a blueprint or definitive template (Jeffrey and Gearey, 2006). Although this book largely focuses on headwaters, water quality and diffuse pollution, the case studies provide an opportunity to assess the factors that initiate and sustain development of a holistic and integrated approach to catchment management, and begin to identify essential commonalities, principles and practicalities.

## How can a catchment management programme get started?

The questions of how best to initiate and evolve a catchment management programme are non-trivial. The complexity and challenges of catchment management raise fundamental questions about governance, and specifically the organizational and institutional structures and allocation of responsibilities best able to develop and implement solutions for a given catchment (within the framework of multi-level governance prevailing in its jurisdiction).

For complex and difficult environmental management problems it is observed that there has been a trend away from monocentric and centralized 'command-and-control' regulation by government, towards relatively decentralized, polycentric and multi-level governance arrangements (see for example, Hardy and Koontz, 2009). These seek to address concerns at an ecosystem or catchment scale and typically depend on both inter-organizational collaboration and action in the state sector, and collaboration between state and non-state actors including citizens. Definitions vary but collaboration can be understood as diverse stakeholders working together to develop and advance a shared vision. It will involve representatives of a wide cross-section of organizations, interest groups and other citizens with a stake in the outcome, and processes aimed at problem-solving and consensus building in pursuit of agreed goals (Benson *et al.*, 2013).

The environmental management literature suggests multiple approaches for assessing collaboration, including a focus on actors, processes, institutions and scales. Of particular interest for this book are two aspects. First, the extent of partnership working between government and non-government actors, including private businesses and individuals, as this is indicative of collaborative governance compared to 'command-and-control' government. Second, the degree to which responsibility for catchment management is delegated to the local or catchment scale and how this is achieved, connecting to debates over the appropriate role of government vis-à-vis private actors in environmental management (Koontz *et al.*, 2004) and other forms of government decision-making (O'Leary and Blomgren Bingham, 2009).

Collaboration and subsidiarity (of decision-making) in environmental management have been deliberately fostered by supra-national and national conventions and legislation. For example, Principle 10 of the 1992 United

Nations Rio Declaration and Section III of the UN Agenda 21 stress that public participation in decision-making is essential for achieving sustainable development. Together, these principles were formative for the 1998 UN Aarhus Convention on access to information, public participation in decision-making and access to justice in environmental matters. More specifically for water management the UN 1992 Dublin Principles and Chapter 18 of Agenda 21 propose that public participation is essential for effective IWRM. At national level collaboration has been deliberately fostered via instruments such as the EU WFD, the Australian Natural Heritage Trust and the US Clean Water Act (CWA) (Benson *et al.*, 2012).

What is also evident, however, is that collaborative approaches have also emerged to an equal (or even greater) degree through grassroots responses to specific local environmental problems; that is, there is self-organizing governance (Benson *et al.*, 2013). As noted above, there are examples of catchment management programmes that have originated in this way. In such cases new forms of governance at a local level find ways to interact and if possible assimilate with government, but the existing literature remains largely unclear on exactly how, why and with what overall effect this takes place (Benson *et al.*, 2013). Thus the case studies in this book aim to provide an opportunity for comparison of mechanisms by which collaborative and decentralized governance arrangements for catchment management can commence and evolve.

## What approaches and tools are needed for catchment management?

In Chapter 1 discussion of the challenges of catchment management demonstrated that it is a complex and difficult area for public policy and government. The challenge of diffuse water pollution in particular exhibits bio-physical processes that may not be directly observable (at least until severe negative impacts become apparent), technical uncertainty regarding the underlying science of the sources, pathways, receptors and impacts of pollution, and societal uncertainty concerning goals, values and allocation of the costs of improvement. Catchment management inevitably involves issues which are contested and over which there may be conflict of interests (Smith and Porter, 2009). It can thus be expected that a catchment management programme will require solid scientific credentials and that its assessments and recommendations should stand up to technical scrutiny and have credibility with all stakeholders. The case studies in this book provide an opportunity to learn how such a scientific and knowledge-sharing agenda has been successfully addressed; indeed to learn how science and governance have been integrated for catchment management.

Further to this, the uncertainty, complexity and dynamics of the catchment management challenge also suggest that it is likely to be impossible to comprehensively determine and implement a solution to the 'problem' through a single sequential approach of problem diagnosis, planning and implementation (Smith and Porter, 2009). Adaptive management is a way to accommodate the uncertainty inevitable in environmental management and is a logical, systematic process for improving management by learning from the outcomes of policies and practices that have been implemented. Adaptive management can be defined narrowly as a rigorous approach of management experimentation to learn how a system responds (Walters, 1986; Walters and Holling, 1990). However, it has become widely understood as a more broadly defined iterative process of assessing system performance, implementing management actions and then monitoring and learning from outcomes (Gregory *et al.*, 2006; Pahl-Wostl, 2008). Such adaptive capability needs to be established with a focus on process but without making this an end in itself, or losing sight of goals for improved socio-economic or environmental outcomes (Pahl-Wostl *et al.*, 2007).

Social learning, understood as the capability to transform a problem situation through a change in understanding and practice (Ison and Collins, 2008), is an emergent property of collaborative adaptive management. Social learning in catchment management is expected to incorporate co-creation of knowledge with stakeholders, change in stakeholder perceptions and values over time, strengthening of analytical capacity, building of social capital and trust between partners, and change in behaviour by individuals and organizations (Smith and Porter, 2009).

Although it presents a logical strategy, an adaptive approach supported by social learning can be constrained by existing social and institutional factors (Allan *et al.*, 2008). These can include bureaucratic fear of failure and of being held accountable, inadequate protocols for monitoring and evaluation of outcomes, and inadequate resources and political will to allow maturity of the cycle of action and reflection. Existing catchment planners and managers may require training, and administrative and political support, to implement adaptive management. The case studies in this book provide an opportunity to see how an adaptive approach has been applied in catchment management, whether and how social learning has occurred and how challenges such as those identified in this paragraph have been overcome.

## How do things get done?

It has been observed so far that the complex and difficult nature of the catchment management challenge stands to require a catchment-based, collaborative and adaptive approach well-matched with the aims and regulations of higher-level policy and governance. How then to sustain and develop such arrangements once they have emerged? How then to sustain integrated and catchment-based assessment, planning and implementation of measures so that water quality and other environmental and socio-economic goals are achieved?

Does this require an appropriate combination of the 'top-down' and the 'bottom-up' given the realities of multi-level governance that prevail? If so, how can sufficient vertical coordination of roles, responsibilities, actions, information flows and resources be achieved? How can implementation at the catchment scale by multiple organizations be coordinated horizontally, and how can the necessary partnership arrangements for this be given sufficient standing and legitimacy to have both local acceptance and be well-assimilated in existing multi-level governance structures? What other characteristics are essential for local-level catchment organizations or partnerships to succeed? What forms and strengths of leadership are required? How can catchment management tasks that necessitate action across administrative boundaries, and at a scale sufficient to account for physical and socio-economic interdependencies and scale effects, be carried out?

The case studies in this book provide an empirically grounded opportunity to uncover at least some of the answers to this very demanding range of vital questions. The challenge is to identify principles and commonalities in how effective collaborative and multi-level partnerships for catchment management can be built, should operate and be sustained, and the essential political, legislative, administrative, scientific and other provisions necessary to support this.

## What are the outcomes of catchment management?

Evaluation of the processes and outcomes of catchment management programmes is necessary not only for evaluating effectiveness in terms of natural resource management and sustainability, but also for the social learning and reflexivity inherent in an adaptive management approach. This chapter has set out the expectation that catchment-based, collaborative and adaptive working should be able to integrate and manage multiple stakeholder interests and knowledge, build trust and social learning, develop solutions for which there is consensus, and lead to improved environmental and socio-economic outcomes despite the complex challenges of catchment management. The case studies in this book illustrate the potential for all this to be achievable.

In the literature, frameworks and empirical evidence for evaluation of the effectiveness of collaborative catchment processes and their outcomes are considered poorly developed and limited (Muñoz-Erickson *et al.*, 2010). However, a wide range of evaluation criteria have been proposed. These can be best classified into: process criteria, environmental outcome criteria and socio-economic outcome criteria (Conley and Moote, 2003). Criteria for environmental and socio-economic outcomes should potentially encompass the unanticipated as well as the targeted and planned, although provision for monitoring will inevitably be limited in scope by resource constraints. It should also be noted that key measures of outcomes will be set by the regulatory objectives or standards in a particular jurisdiction. For example, under the federal Clean Water Act in the USA waterbodies are assessed in terms of their designated uses such as water supply, fisheries, recreation and navigation. From a legal and regulatory perspective the key measure of outcomes thus becomes whether each waterbody is sustained in its particular designated use.

Significant challenges exist in applying environmental outcome criteria to catchment management initiatives: the problem of identifying measurable goals; the variability inherent in environmental data, which combined with the long timeframe required for changes to occur makes identifying trends difficult; the typical lack of long-term data from a carefully designed and adequately resourced monitoring programme; and the attribution challenge, as making causal links between specific management activities and environmental trends is often not possible when it is difficult to isolate variables. Ultimately, rigorous evaluation requires baseline data and regular monitoring of key indicators. Similar if not greater challenges arise in trying to evaluate socio-economic outcomes such as community well-being or farm business viability (Conley and Moote, 2003). Despite these challenges an attempt has been made to evaluate and summarize the outcomes to date of each of the case studies in the chapters that follow.

## Selection and presentation of the case studies

The description of each case study was compiled and critically reviewed by a team of principals representing each catchment. Subject to the limits of the available information an attempt has been made to characterize and evaluate each case study catchment according to consistent criteria informed by the premises identified in Chapter 1 and in this chapter, and using a common narrative framework based on the five key questions (and sub-questions) identified at the beginning of this chapter.

This chapter concludes with a brief introduction to the case studies selected and the reasons for their selection. The Upper Susquehanna River Basin (Chapter 3) constitutes the headwaters of the Chesapeake Bay watershed programme. Established under the federal Clean Water Act, this programme has the highest priority of restoring and protecting the greatly valued resources of the Chesapeake Bay. The Upper Susquehanna Coalition (USC) is a network of county-level natural resource professionals formed to develop strategies, partnerships and projects to protect the headwaters of the Susquehanna River and thus the upper Chesapeake Bay watershed. It provides an example of a mainly publicly funded yet 'bottom-up' network that has thrived despite a relative lack of formal legal status through innovative local leadership, strong technical capacity and the establishment of trust and goodwill with partner agencies and wider stakeholders.

In the USA the New York City Watershed Protection Progam (Chapter 4) is recognized as being the most significant attempt to protect at its watershed source the water supply of a major metropolitan region. The surface waters from the Catskill and Delaware watershed that supply the city carry little sediment and are not filtered before delivery to consumers. Through a Watershed Agreement with federal, state and local stakeholders the city aims to maintain this high-quality supply and to manage conflict between economic growth in the watersheds and deterioration in water quality. Measures for water quality protection are partly voluntary, partly regulatory and fully funded by the city, but are managed by local authorities and institutions. The case is often cited as an example of a 'payments for ecosystems services' scheme (the water consumer paying the upper watershed for clean water supply), but more importantly it is a test case for the premise that local capacities are the best means to meet water quality goals in a watershed. Despite success in water protection, tensions remain between the city and watershed communities concerning the scale and pattern of protection measures and the perceived trade-offs between economic growth and environmental protection.

The Hudson River Valley (Chapter 5) is one of 40 National Heritage Areas designated by the United States Congress. The Hudson River Estuary Program (HREP) leads a unique regional partnership to restore the Hudson through clean-up of pollution, conservation of natural resources and the quality of life of basin communities, and promotion of public use and enjoyment of the river. HREP provides an example of an organization that although part-funded by New York State and based in the state's Department for Environmental Conservation is neither a regulatory agency nor a river-basin-based regional authority. Part funder and part facilitator of action, it is a pragmatic coordinator of partnerships that leads development of strategic plans to be delivered by facilitated and coordinated action by others. Partners include local governments, non-governmental organizations, environmental groups, utilities, businesses, researchers, educators, recreational groups and other concerned citizens. Sustained efforts to serve the multiple interests of people in the watershed has resulted in actions with widespread acceptance and achievements ranging across improvements in water quality, enhanced and protected landscapes and capacity building.

The South East Queensland (SEQ) Healthy Waterways Partnership (HWP) facilitates collaborative water resource management across 19 catchments centred on the Brisbane River and Moreton Bay (Chapter 6). From its origins as a scientific study conducted by concerned local actors, the HWP has evolved over more than 20 years into a membership network of local governments, state and federal partner organizations, non-governmental organizations, environmental groups, community groups, industry partners and scientists. The HWP has a shared vision for 'healthy waterways' and socio-economic development that facilitates collaboration to deliver agreed strategic plans and action agendas. With some similarities to the HREP it illustrates how a network can achieve scientific, spatial, collaborative and operational integration, and a holistic approach. Success is demonstrated by well-developed monitoring but the developmental pressure in one of Australia's fastest-growing regions and extreme weather events present continuing challenges for the health of the region's waterways.

The Drastrup Project of the City of Aalborg, Denmark (Chapter 7) has protected groundwater and the city's water supply through land use change and the removal of polluting activities from designated areas. It illustrates how this can be achieved through a unified approach to scientific assessment, spatial planning, stakeholder engagement and partnership working.

The Groundwater Protection Programme of the Water Board of Oldenburg and East Frisia (OOWV) in north-west Germany (Chapter 7) also demonstrates how water supplies from groundwater can be protected. OOWV and local government developed a comprehensive programme of groundwater protection based on cooperation between local partners. The main elements of this programme are regulation, cooperation with farmers, promotion of organic farming and purchase of land for afforestation; all supported by scientific research and public outreach.

Groundwater protection in Drenthe Province of the Netherlands (Chapter 7) has focused on the multiple benefits of floodplain restoration for flood alleviation, wetland and biodiversity conservation and recreation, as well as control of nitrate and other pollution to protect drinking water supply. Achieving the desired patterns of land use and management has required cooperation between stakeholders including landowners, the drinking water utility, the Water Board, the municipality, the provincial government and environmental organizations. The programme provides a further illustration of the benefits of integrated spatial planning, but also shows that success depends on broad-based public acceptance and support.

The Ballinderry River Enhancement Association (BREA) RIPPLE Project, Ulster, Northern Ireland (Chapter 8), has demonstrated a consistently collaborative and inclusive approach and represents one of the few catchment-wide programmes in the UK where members of the local community have exercised genuine power over the decision-making process and played a leading role in project delivery. The importance of trust, social capital and social learning at a catchment scale is demonstrated. The project is aiming to deliver multiple benefits and has created a shared responsibility for the management of the river across the community and partner organizations. However, despite the growing empowerment of the local community, project funding has been limited and more secure institutional standing and finance may be required in future.

Chapter 9 tells the story of the Loweswater Care Programme in north-west England and how it evolved through the interaction between community initiatives and university-based researchers. Loweswater is a lake and its catchment is relatively small in scale compared to the other cases in this book. This case provides an interesting opportunity to focus on how scientific research was integrated with stakeholder engagement and local partnership working to achieve collective knowledge generation and sharing. The lessons that emerge from this local-level example are related in the chapter to regional and national policies and programmes in England, but are also set in context by comparison to the other cases presented in this book.

## Note

1   IWRM is related to the concept of integrated catchment management but is a broader and less place-specific process. IWRM seeks to shift catchments from sector-based

management of water in which organizational and spatial fragmentation is apparent to institutionalized cooperation between organizations and comprehensive multi-level governance. IWRM envisages more coordinated development and management of land and water, surface and groundwater, and upstream and downstream interests. It implies integrated management of water across sectors (energy, fisheries, water supply and sanitation, agriculture, tourism and industry), and across water management functions (water allocation, pollution control, monitoring, financial management, flood management, drought management, information management, basin planning and stakeholder participation). The IWRM paradigm emphasizes three areas of intervention: an enabling policy environment, organizational capacity for implementation, and management instruments such as plans, approaches for conflict resolution, water demand management and regulations.

# References

Allan, C., Curtis, A., Stankey, G. and Shindler, B. (2008) 'Adaptive management and watersheds: A social science perspective', *Journal of the American Water Resources Association*, vol. 44, no. 1, pp. 166–174.

Benson, D., Jordan, A. and Huitema, D. (2012) 'Involving the public in catchment management: An analysis of the scope for learning lessons from abroad', *Environmental Policy and Governance*, vol. 22, no. 1, pp. 42–54.

Benson, D., Jordan, A. and Smith, L. (2013) 'Is environmental management really more collaborative? A comparative analysis of putative "paradigm shifts" in Europe, Australia, and the United States', *Environment and Planning A*, vol. 45, no. 7, pp. 1695–1712.

Conley, A. and Moote, M. A. (2003) 'Evaluating collaborative natural resource management', *Society and Natural Resources*, vol. 16, pp. 371–386.

Gregory, R., Failing, L. and Higgins, P. (2006) 'Adaptive management and environmental decision making: A case study application to water use planning', *Ecological Economics*, vol. 58, no. 2, pp. 434–447.

GWP (2000) 'Integrated water resources management', *TAC Background Papers No. 4*, Technical Advisory Committee, Global Water Partnership, Stockholm.

Hardy, S. D. and Koontz, T. M. (2009) 'Rules for collaboration: Institutional analysis of group membership and levels of action in watershed partnerships', *The Policy Studies Journal*, vol. 37, no. 3, pp. 393–414.

Ison, R. and Collins, K. (2008) 'Public policy that does the right thing rather than the wrong thing righter', Analysing Collaborative and Deliberative Forms of Governance Workshop, 14 November 2008, The Australian National University, Canberra.

Jeffrey, P. and Gearey, M. (2006) 'Integrated water resources management: Lost on the road from ambition to realisation?' *Water Science and Technology*, vol. 53, no. 1, pp. 1–8.

Koontz, T. M., Steelman, T. A., Carmin, J., Smith Korfmacher, K., Moseley, C. and Thomas, C. W. (2004) *Collaborative Environmental Management: What Roles for Government?* Resources for the Future, RFF Press, Washington, DC.

Mostert, E. (1999) 'Perspectives on river basin management', *Physics and Chemistry of the Earth, Part B: Hydrology, Oceans and Atmosphere*, vol. 24, no. 6, pp. 563–569.

Muñoz-Erickson, T. A., Aguilar-González, B., Loeser, M. R. and Sisk, T. D. (2010) 'A framework to evaluate ecological and social outcomes of collaborative management: Lessons from implementation with a northern Arizona collaborative group', *Environmental Management*, vol. 45, no. 1, pp. 132–144.

Murray–Darling Basin Ministerial Council (2001) *Integrated Catchment Management in the Murray–Darling Basin 2001–2010*, Murray–Darling Basin Ministerial Council, Canberra.

O'Leary, R. and Blomgren Bingham, L. (2009) *The Collaborative Public Manager: New Ideas for the Twenty-first Century*, Georgetown University Press, Washington, DC.

Pahl-Wostl, C. (2008) 'Requirements for adaptive water management' in Pahl-Wostl, C., Kabat, P. and Moltgen, J. (eds) *Adaptive and Integrated Water Management: Coping with Complexity and Uncertainty*, Springer-Verlag, Berlin, pp. 1–22.

Pahl-Wostl, C., Sendzimir, J., Jeffrey, P., Aerts, J., Berkamp, G. and Cross, K. (2007) 'Managing change toward adaptive water management through social learning', *Ecology and Society*, vol. 12, no. 2, art 30.

Rockström, J., Falkenmark, M., Folke, C., Lannerstad, M., Barron, J., Enfors, E., Gordon, L., Heinke, J., Hoff, H. and Pahl-Wostl, C. (2014) *Water Resilience for Human Prosperity*, Cambridge University Press, Cambridge.

Smith, L. E. D. and Porter, K. S. (2009) 'Management of catchments for the protection of water resources: Drawing on the New York City watershed experience', *Regional Environmental Change*, vol. 10, no. 4, pp. 311–326.

Walters, C. (1986) *Adaptive Management of Renewable Resources*, McGraw Hill, New York.

Walters, C. J. and Holling, C. S. (1990) 'Large-scale management experiments and learning by doing', *Ecology*, vol. 71, no. 6, pp. 2060–2068.

WWF (2013) *Integrated River Basin Management (IRBM): A Holistic Approach*, www.wwf.panda.org/about_our_earth/about_freshwater/rivers/irbm/, accessed 29 August 2013.

# Part II
# Case studies

# 3 The Upper Susquehanna River Basin

Headwaters of a national treasure
– the Chesapeake Bay

*Keith Porter, James Curatolo, Mike Lovegreen
and Laurence Smith*

## Introduction

On May 12, 2009, President Obama issued Executive Order 13508. Its declared purpose was "to protect and restore the health, heritage, natural resources, and social and economic value of the nation's largest estuarine ecosystem and the natural sustainability of its watershed." The Order affirmed that the "Chesapeake Bay is a national treasure" as the largest estuary in the United States. It is also the third largest in the world and one of its most ecologically productive estuaries. Although the Order clearly established a strong leadership role for the federal government, it also recognized that success is contingent on collaboration with state and local governments.

In response, the US Environmental Protection Agency established the Chesapeake Bay Watershed Total Maximum Daily Load (TMDL). EPA characterizes the TMDL as being a historic and comprehensive "pollution diet." Each state in the watershed is expected to play its part in meeting this "diet." For New York State, where the northern headwaters of the Chesapeake Bay lie, there is a special challenge. The Upper Susquehanna River is generally of very good water quality. This means that further improvements in what is already of good quality will not be easily and inexpensively accomplished. It will require a comprehensive and sustainable strategy for all significant sources of nutrients and sediment.

Leading the delivery of this strategy "on the ground" is the Upper Susquehanna Coalition (USC, 2013), a network of 16 Soil and Water Conservation Districts (SWCDs) in New York State and three SWCDs in Pennsylvania. Their mission is to protect and improve water quality and natural resources in the Upper Susquehanna River Basin with the involvement of citizens and agencies through education, partnerships, planning, implementation and advocacy. Their overarching goal is that the river leaving the headwaters is as clean as possible, passing the water's stewardship to the downstream states.

Each state in the Susquehanna Basin, including New York and Pennsylvania, has been assigned a load allocation to meet for sediment, nitrogen and phosphorus. The New York Department of Environmental Conservation in partnership with the Upper Susquehanna Coalition (USC) has developed a local county-based Watershed Implementation Plan to reduce these loads as the state's contribution to clean up the Bay. The state's approach is to target the highest quality practices that will help reduce its nutrients and sediment loads regardless of the geographical location. This unique approach is possible because the USC has a watershed-wide network capable to implement projects. The USC's Pennsylvania members are also working closely with their state's Department of Environmental Protection to implement that state's watershed improvement strategy.

## Context and drivers

### *The Chesapeake Bay problem*

The Chesapeake Bay is an extremely diverse ecosystem and the most studied estuary in the United States. Such scrutiny is motivated by the Bay's scale, its beauty and economic value. Approximately 322 kilometers in length, and varying in width from 5.5 to 56 kilometers, the Bay is relatively shallow for its size, with an average depth of only 6.4 meters. With a shoreline 19,000 kilometers in length, greater than the entire west coast of the United States, it forms a critical part of the mid-Atlantic coastal environs, and receives freshwater from a catchment that is almost 166,000 square kilometers in area, extending from New York State in the north to Virginia in the south. This watershed encompasses parts of six states: Delaware, Maryland, New York, Pennsylvania, Virginia, West Virginia, and also includes the District of Columbia. The watershed is inhabited by a population of about 16 million people.[1]

Until the last few decades the Bay supported richly abundant plant and animal life, and has been termed "an immense protein factory" (Davison *et al.*, 1997). Its name Chesapeake itself originates from a Native American word "Tschiswapeki" meaning "great shell fish bay." That copiousness supported very heavy fishing. Unfortunately the fishing, together with an increasing density of human population in the surrounding regions, has caused serious deterioration in the waters of the Bay and its ecosystem. A particularly damaging consequence of the deterioration in water quality is hypoxia. This life-inhibiting condition occurs when the decay of excess amounts of plankton or algal biomass causes the concentration of dissolved oxygen to fall below two parts per million (Smith *et al.*, 1992). High levels of plankton or algal blooms are in turn the result of elevated inputs of nutrients to the Bay, especially nitrogen. Corresponding to increasing nutrient inputs due to population growth since the 1950s, the hypoxic volume in the Bay has increased more than fourfold, exceeding 10 percent of the Bay's waters (Horton, 2003). This creates what Horton terms an underwater desert, and hypoxia is a primary measure of the health of the Bay.

## A restoration campaign

By the mid-twentieth century the degradation of the quality of the Bay caused increasing concerns. Over two decades, the deterioration prompted Congress to fund major studies costing millions of dollars (Davison *et al.*, 1997). In the mid-1970s, US Senator Charles Mathias, a Republican representing Maryland, led a campaign to persuade Congress to provide a further US $27 million for a five-year study. This study concluded that excessive nutrients were the cause of the Bay's decline. A major source of the nutrients was determined to be run-off from farm fields (Davison *et al.*, 1997). In 1983, the governors of Virginia, Maryland, Pennsylvania, the Mayor of the District of Columbia and the US Environmental Protection Agency (USEPA), jointly developed a "Chesapeake Bay Agreement" with a purpose of overseeing the implementation of coordinated plans to improve and protect the water quality and living resources of the Chesapeake Bay estuarine systems. Four years later, the campaign begun by Senator Mathias culminated in a Congressional amendment of the Clean Water Act creating the USEPA Chesapeake Bay Program Office (Clean Water Act, 33 USC §1267 Chesapeake Bay).

The new Chesapeake Bay Program Office announced the goal of reducing the loads of nitrogen and phosphorus entering the Bay by 40 percent by 2000. Unfortunately, this goal was defeated by the increasing density of land uses and growth in population throughout the period 1985–2000. Rather than improvement, extensive monitoring of water quality in the Bay by the US Geological Survey showed generally either non-significant or modest downward trends in nutrient and sediment loads, and concentrations (Langland *et al.*, 2001). This insufficient restoration of the Bay over several decades finally prompted USEPA to take regulatory action.

## Total maximum daily loads or caps

Under the federal Clean Water Act, if a contaminant reaches a level that impairs designated uses of the water body, the responsible regulatory agency must estimate and enforce the maximum level of the contaminant that safely allows those designated uses. This maximum level is termed the Total Maximum Daily Load or TMDL. A TMDL is a load "established at a level necessary to implement the applicable water quality standards with seasonal variations and a margin of safety which takes into account any lack of knowledge concerning the relationship between effluent limitations and water quality" (Clean Water Act, 33 USC §1313(d)(1)(C)).

In the case of the Chesapeake Bay, the USEPA for many years failed to require the imposition of TMDLs for nutrients or sediments, the contaminants determined to be the cause of the impairment of the Bay. Six lawsuits were filed against the USEPA by concerned citizens in the watershed states, leading to the outcome that the USEPA signed consent decrees setting out schedules for establishing TMDLs. However, their implementation was temporarily deferred because in June 2000 the Chesapeake Bay Program partners signed the Chesapeake Bay 2000 Agreement. Under this agreement the partners committed themselves to "nurture and sustain" a Chesapeake Bay Watershed Partnership and to achieve goals and a schedule of actions outlined in the agreement. More importantly, the agreement led to the specification of a voluntary form of TMDL in the form of cap loads of nitrogen, phosphorus and sediment. This action deferred the formal application of TMDL regulations by the USEPA.

Initially the Chesapeake Bay Program partners agreed to limit annual nitrogen loading to the Bay at 175 million pounds (80 million kilograms) and annual phosphorus loading at 12.8 million pounds (5.8 million kilograms). These loading limits compare with watershed model estimates of the actual levels of nitrogen and phosphorus in 2000 of 285 million pounds and 19.1 million pounds respectively. Thus annual reductions had to be sought of about 110 million pounds of nitrogen (50 million kilograms) and 6.3 million pounds of phosphorus (2.9 million kilograms). Excessive water-borne sediment also reduces the light necessary to support healthy aquatic life and can be deposited as a smothering blanket. The Bay mathematical model calculated that the sediment loads that would not compromise desired levels of water clarity were 4.15 million tons, a loading of 0.89 million tons per year less than the 2000 estimated cap load of 5.04 million tons (1 US ton equals 0.91 metric tonne).

Cap loads are in fact equivalent to TMDLs less regulatory enforcement. They allocate load reductions to sources—point, non-point and atmospheric—just as a TMDL but on a voluntary basis (Porter, 2004; Porter, 2010). A significant expression of dissatisfaction with this deferral of regulations was the Petition of the Chesapeake Bay Foundation (CBF), an influential non-governmental organization of significant membership, professional staff and funding. In December 2003, the CBF petitioned the USEPA under the Administrative Procedures Act 5 USC §553(e), "to issue, amend, or repeal rules and take corrective action relating to the regulation, control, and permitting of point source discharges of nutrients (nitrogen and phosphorus) from significant sewage and industrial treatment plants in the Chesapeake Bay Watershed."

The lack of confidence expressed by the Chesapeake Bay Foundation in 2003 was confirmed by the continuing poor quality of the Chesapeake Bay. In its assessment of the Bay in 2008, the USEPA's Chesapeake Bay Program admitted that:

> Despite small successes in certain parts of the ecosystem and specific geographic areas, the overall health of the Chesapeake Bay did not improve in 2008. The Bay continues to have poor water quality, degraded habitats and low populations of many species of fish and shellfish. Based on these three areas, the overall health averaged 38 percent, with 100 percent representing a fully restored ecosystem.

(USEPA, 2008, p. 4)

Since 2007, the University of Maryland Center for Environmental Science (UMCES) has published a Chesapeake Bay Health Report Card. The Card provides a scientifically based and geographically detailed annual assessment of the conditions of the Bay. The report card provides a rating of 15 regions in the Bay. The rating is based on six indicators, three measuring water quality and three biotic measures. These six indicators are combined into an integrated indicator of health. Using these comprehensive measures, the UMCES gave the Chesapeake Bay an overall grade of C– in both 2007 and 2008, and a grade of C in 2009. Regrettably in 2010 the overall health of the Bay declined for the first time in four years. Ratings for four of the six indicators—chlorophyll *a*, water clarity, aquatic grasses and benthic community—all declined according to the assessment, dissolved oxygen remained steady and phytoplankton community scores slightly increased (UMCES, 2010). For 2012 and also for 2013, the report card indicates a small improvement in the health of the Bay sufficient to be given the grade C. This is up from the D+ given by the 2011 report (UMCES, 2012; UMCES, 2013).

It was the failure of the restoration programme that prompted President Obama's Executive Order. The primary outcome of the Order is "a major initiative to establish and oversee achievement of a strict 'pollution diet' to restore the Chesapeake Bay and its network of local rivers, streams and creeks" (USEPA, 2010, p. 1). A "Bay TMDL" and accompanying implementation plans will accomplish this "diet." The Bay TMDL will be a combination of over 90 smaller TMDLs for individual tidal segments in the Bay. The "diet" derived from these combined TMDLs will then be divided and allocated by USEPA among all the jurisdictions in the watershed: Maryland, Virginia, Pennsylvania, Delaware, New York, West Virginia and the District of Columbia. These jurisdictions will then allocate their respective target loadings of nutrients and sediment within their respective river basins.

### The challenge for New York State and the counties of the Chesapeake Bay headwaters

USEPA has declared that the draft allocations for New York State are almost 4 million kilograms of nitrogen and almost 260,000 kilograms of phosphorus. These allocations compare, for example, with an estimated existing load of nitrogen of approximately 4.8 million kilograms. As noted in the introduction to this chapter the imposition of TMDLs is challenging for New York State because surface waters in the Upper Susquehanna are generally of very good quality. The Susquehanna River Basin in New York consists of 7,597 miles (12,225 kilometers) of rivers and streams and over 400 lakes and ponds. Of these over 85 percent fully support their designated uses with only minor impairments from pollution. Uses are impaired in only about 10 percent of basin waters, but such impairments are almost entirely due to mercury in the waters leading to them being under a "fish consumption advisory." Fish consumption advisories are issued routinely by the New York State Department of Health to recommend limits on the consumption of sports fish according to the level of the contaminant where the fish are caught. The origin of the mercury is largely attributed to atmospheric deposition. One reservoir in the Upper Susquehanna River Basin is also designated as impacted by farming activities. Lesser impacts or threats affect another 10 percent of basin waters in the watershed (NYSDEC, 2009). The most significant sources of these minor impairments include agricultural activities, inadequate on-site septic systems and stream bank erosion. A major tributary of the Susquehanna in the state, the Chemung River Basin, has a good quality that also meets it designated uses throughout the vast majority of its river segments (NYSDEC, 2007). In summary, the Chesapeake Bay Program is challenging New York State to make clean water cleaner!

This might be considered sufficient reason for objection, but further controversy has surrounded the technical basis for load allocations imposed on New York State. Many millions of dollars have been invested in scientific studies of the Chesapeake Bay and to a lesser extent its watershed. This investment includes development of the mathematical model of the watershed funded by USEPA. The model used is a complex version of the Hydrological Simulation Program–Fortran (HSPF) model. The results of this model for the Susquehanna River Basin in New York State are questioned by the New York State Department of Environmental Conservation and the USC who do not regard the model as sufficiently accurate despite the huge investment in it. This is crucial given that the model results provide the basis for the TMDL allocations. The model bases its New York State calculations on data from a monitoring station at Towanda, located in Pennsylvania and some distance south of New York State. This anomalous application of the model reflects the lack of a continuous monitoring station for water quality in the New York State part of the Chesapeake Bay Watershed. The model is also criticized for its reliance on assumptions and purely mathematical and unvalidated estimates of the loads and from where they originate.

A number of further issues combine to make this an exceptional case study of the application of TMDLs as a mechanism to control diffuse pollution in rural areas. First is that the beneficiaries of New York State successfully meeting goals set under the Chesapeake Bay Program will reside in downstream states. New York is being charged by the USEPA to improve already good quality waters mainly for the benefit of downstream neighbors and the Chesapeake Bay, located more than 300 kilometers south of New York State. As one New York farmer complained: "Why should we make good quality water better for rich people yachting in the Chesapeake Bay?"

New York State was a late signatory to the Chesapeake Bay Program, only becoming an official partner when it signed the Chesapeake 2000 Agreement. There is a feeling in the state that most of the federal investment has been made in downstream states, and as a headwater state New York has been somewhat neglected. This is manifest in the inadequate monitoring of water quality in the Upper Susquehanna River noted above. Despite this, New York State has invested heavily in agricultural and other programmes to reduce both diffuse and point-source pollution; a major reason why water quality in the state's part of the Susquehanna Watershed is so good. Such cannot be claimed for the downstream states, which have more polluted waters.

Even the EPA has stated that if all the waters flowing into the Chesapeake Bay were of equal quality to the Susquehanna in New York State the quality of the Bay would be such that there would be no need for the TMDLs. Indeed, should the Bay meet its TMDL targets, all of the downstream states' waters will still have higher nutrient levels than New York has at present. Hence there is inequity in the costs so far incurred. It would seem to be fairer if the burden of restoring the Chesapeake Bay now fell on the states other than New York. However, the USEPA has determined that the TMDL should be applied not according to water quality but on a simple proportional basis. Thus New York is being required to achieve the same percentage decrease in nutrient and sediment reductions as the other states, regardless of difficulty and the additional cost per unit reduction that this is likely to incur. Apart from the inequity of this arrangement it would clearly be more efficient at a watershed scale to rank all sources of contamination and target load reductions and funds available to achieve these where the biggest improvements can be made.

These inter-state equity issues are significant, not least because approximately 95 percent of the Upper Susquehanna River Basin is rural and its residents are not generally affluent. They are therefore understandably reluctant to adopt, without compensation, costly pollution mitigation measures for the remote needs of those who benefit from improvements to the Chesapeake Bay. Such reluctance is compounded by the fact that most federal and state grants and incentives available to farmers give priority to basins where surface or groundwater bodies are impaired in their designated uses. Such an allocation of grant funds can be seen as favoring those farmers who have not adopted best management practices while denying funds to those who have. Thus "bad actors" are more likely to be rewarded than good custodians of the environment, and in simple practical terms New York may find it more difficult to finance programmes that help its farmers make further improvements.

Despite these issues and challenges, New York State also has advantages. First, there is a strongly held ethic in favor of protecting water quality reflected in state law. Article 11-A of the New York State Agriculture and Markets Law establishes the Agricultural Environmental Management (AEM) programme to assist farmers in managing their farm operations in a way that protects the environment and helps maintain the economic viability of the farm. The AEM programme provides a framework of institutional and technical support to the farming community, and can be delivered by a strong network of technical providers for farmers and rural communities. This institutional and technical support has been especially effective in the Upper Susquehanna River Watershed since the creation of the Upper Susquehanna Coalition.

## Getting started

### Origins of the Upper Susquehanna Coalition

In 1992, managers of several Soil and Water Conservation Districts (SWCDs) representing counties in Pennsylvania and New York met to discuss potential collaboration in the Upper Susquehanna River Basin. Their purpose was to explore creating some new entity capable of providing watershed-wide coordination and leadership. The outcome was the establishment of the Upper Susquehanna Coalition, with its declared purpose to develop strategies, partnerships, programmes and projects to protect the headwaters of the Susquehanna River in the Chesapeake Bay Watershed. Originally 11 counties in New York and three in Pennsylvania formed the USC; another five counties in New York State have since joined. Initially the USC functioned entirely on a voluntary basis founded on the trust and goodwill between the SWCDs comprising its core membership, and in 1996 it obtained the funds to hire a full-time coordinator. However, its increasing success in acquiring funds and associated increased administrative responsibilities compelled consideration of a more formal organization.

In 2005, the USC requested Eric Horan, a student in the Cornell Law School Water Law Clinic, to assess legal instruments that could form the basis of a readily acceptable agreement (Horan, 2005). Horan recommended that a Memorandum of Understanding (MOU) between the SWCD members would be least constraining while having some legal weight, given that this is a commonly used legal tool to formalize a partnership between two or more organizations. While MOUs have varying levels of commitment, most outline matters of general agreement between participating parties for a set period of time or a given subject matter. At the request of the USC Horan developed an outline for a MOU that would be sufficiently robust for the districts to work together and also acceptable to the states of New York and Pennsylvania. The draft then written by the USC coordinator ultimately led to a MOU that was accepted and signed by all the counties comprising the USC and both state's Conservation Committees within their Departments of Agriculture. The MOU unites its members in a genuinely coordinated effort and to date has maintained remarkable cordiality and cooperation between its members and partners.

### Soil and water conservation districts: a local resource for environmental management

The origin of the soil and water conservation districts lies in the early 1930s. Following sustained dry weather conditions across the United States, dried out soil in the heartlands of the USA began to erode. Winds created huge dust storms that obliterated sunlight and engulfed the land. The storms stretched south to Texas and north-east to New York. This unnatural disaster prompted the US Congress to unanimously pass legislation making soil and water conservation a national policy and priority. Congress fully recognized that the nature of private ownership of land in the US made the success of conservation on private property dependent upon voluntary participation of the landowners in the work. In 1937, President Roosevelt recommended to the governors of all the states that they adopt legislation to establish soil conservation districts with the participation of local landowners. Subsequently, district-enabling legislation was passed in every state. Today, there are nearly 3,000 conservation districts across the United States. As Douglas Helms, national historian for the Soil Conservation Service, has stated in relation to SWCDs:

> In a way the system of district and state cooperation with the federal government could produce a service that was greater than the sum of its parts. For instance, the Soil Conservation Service had the staff to develop standards for the various conservation practices and modify them to fit the local area. But the state, county or districts could accelerate conservation by helping to pay for installing conservation practices or by hiring additional technical staff. In those states which chose to hire additional staff, one might walk into a field and find people paid by the federal government, the state, or the district. Yet all would be doing similar work, using similar methods.
>
> (Helms, 1992, p. 300)

In New York State, the legislature enacted the state's Soil and Water Conservation Districts Law in 1940. In its declaration of policy, the purpose of the law is to preserve soil and water resources, broadly defined, and to utilize those resources to enhance the quality of life. The Soil and Water Conservation Districts Law of New York State § 5(1) authorized the legislature of each county to declare by resolution "the county to be a soil and water conservation district for the purpose of effectuating the legislative policy" of the Soil and Water Conservation Districts Law.

Every county in New York State is now a soil and water conservation district. A SWCD board of directors comprising two members of the county board or legislature, two members who are practical farmers, and one "member at large" representing urban, suburban and rural non-farm landownership interests directs each district, as specified under Soil and Water Conservation Districts Law of New York State § 6(1)(a). This composition of the board ensures that the local interests of the county are the basis of the work of the district. SWCD staff members are trained and experienced technicians who are fully qualified to assist the constituencies of their county in meeting soil and water objectives. Their remit and expertise has broadened well beyond the management of soils to encompass all of the natural resource management challenges of their region.[2] Importantly, SWCDs tend to be generally trusted and accepted by farm and non-farm landowners, always adopting in the first instance an independent advisory rather than regulatory stance. They thus provide a capability for local analysis and extension that is an invaluable resource for watershed management, and if it did not exist might be very difficult to create from "scratch" given the budgetary stringencies of today. As Hugh Hammond Bennett, a "founder" of the conservation district movement, stated:

> One of the best, and certainly the most promising, of the devices yet invented by man for dealing democratically and efficiently with the maladjustments in land use, as well as for carrying forward positive programs of desirable conservation, and for maintaining the work, is the soil conservation district.
>
> (Bennett, 1943, p. 194)

## Approach and tools: the USC integrated watershed approach

The USC's activities consist of three main components: environmentally and economically sustainable agriculture, stream corridor rehabilitation (which includes management of highway drainage ditches and urban and rural stormwater) and wetland restoration. Each of these components, and the USC's contributions to strategic planning for the Chesapeake Bay with the New York State Department of Environmental Conservation, is underpinned by strong scientific capability developed "in house" by USC members and in research partnerships with regional universities. For example, the USC has close working ties with scientists and faculty in the State University of New York at Binghamton and Oneonta, the State University of New York College of Environmental Science and Forestry, and Cornell University. Work has included evaluation of riparian buffer nutrient reduction, "fingerprinting" sediment to determine relative loading rates from landscapes, developing protocols to assess stream banks for long-term stability and erosion potential, and analyzing the landscape to determine where wetland restoration would be most valuable.

The USC's scientific capability includes modeling to provide support for the understanding of nutrient and sediment fluxes throughout the basin and plan implementation of control measures, monitoring to determine water quality and track the effectiveness of implementation efforts, and evaluation of best management practices (BMPs) to determine the best mix of implementation efforts to meet state and county objectives (using criteria of cost-effectiveness, provision of multiple benefits, and accord with local and state needs).

### *Environmentally and economically sustainable agriculture*

The largest component in the USC's activities, this programme involves documenting farm statistics and BMPs, developing watershed and site-specific agricultural plans, and implementing and evaluating practices. The USC uses the New York State AEM programme approach to evaluate farms and follow the Chesapeake Bay Program's approach for best management practice information to determine the status of nutrient loading from the agricultural sector. A significant part of the USC success is the members' and coalition's ability to leverage local, state and federal resources for implementation of BMPs from a variety of programmes and agencies; a role for which SWCDs were specifically created. The USC has also derived direct benefits from the very substantial scientific work completed in the New York City Watershed which is immediately adjacent to the Susquehanna River Basin. An example of this transfer of scientific and technical understanding is the precision-feeding programme in animal production.

The USC emphasizes working with farmers to adopt sound management practices that recognize the economic constraints of the farm as a business while meeting environmental objectives. Principal management measures include conservation planning, conservation tillage, animal waste management systems, nutrient management plans, prescribed grazing systems, and the protection of stream corridors through riparian buffer strips, and fencing to exclude animals from gaining access to the streams.

Farm stewardship has been promoted through a Riparian Buffer Initiative and currently through a major grazing initiative and promotion of nitrogen-reduction strategies. Combining economic incentives with environmental benefits, these initiatives target farms and those sub-watersheds where gains from nitrogen reduction and the creation of effective riparian buffers are most beneficial. The target farms were contacted, project sites reviewed and the best sites chosen using the USC farm inventory database and technical knowledge of the USC grazing specialists. This approach continues through diverse funding from grants, dedicated federal programmes and even environmental fines.

Application of management practices and monitoring are particular attributes of the USC approach rather than investment in soil and water control structures. This provides for a regional approach that directly addresses water quality, stream sediment and farm viability issues. For example, well-managed pasture systems not only provide feed for livestock and wildlife, retain valuable topsoil and improve water infiltration, but are also aesthetically pleasing to residents of the watershed. However, support is also provided to member counties who take the lead for farm infrastructure improvement projects such as manure storages or barnyard renovations. A common structural improvement is the construction of drainage around barnyards to prevent run-off of surface waters contaminated by manure from the barnyard area.

### Stream corridor rehabilitation and emergency stream intervention

This component addresses natural stream design, stream stabilization, floodplain enhancement and stream buffer protection related to the management of stream bank erosion and flood risks. The USC has initiated a "Stream Team approach" since 2009 to provide technical support for member counties and training to build their capability. The USC Stream Team also provides intensive training sessions to landowners and municipal officials on stream natural design techniques. This training also includes and integrates road ditch reconstruction and stabilization. The Team visits USC member sites and provides technical analyses of the streams and aids in designing and implementing rehabilitation projects. Guiding principles of the work are as follows.

Stream issues are considered in a systemic manner considering the condition of the whole watershed and impacts within it. As feasible, streams are monitored to determine the rate and status of observed or perceived impairments. As much as possible the objective is to restore the stream to its natural state as opposed to stabilization. Such restoration includes consideration of geomorphic, hydrologic, habitat, water quality, riparian, social and economic values. In practice remedial measures must be pursued pragmatically, recognizing that funding, materials and other resources are limited. Creative and cost-effective approaches to stream restoration are encouraged in management, regulation and physical in-channel works. Of primary importance is the education of landowners, municipal officials, maintenance personnel, land use planners and other citizens in order to effect a cultural change in the management of streams and watersheds. Allied to this, the aim is to achieve local empowerment and capacity-building through education, training and actual experience, and through the use of local designers, contractors and materials suppliers. The experience and understanding learned in the region for stream restoration will be shared and networked more widely, while further research on regional stream system elements is needed to better understand the complexity of the local streams.

*Highway drainage and stormwater*

Effective stormwater management in the upper Chesapeake Bay Watershed must take account of its unique geology, topography and hydrology that make it one of the most flood-prone regions in the United States. The region's combination of shallow soils, a relatively impervious sub-surface layer or fragipan and hilly topography result in accelerated horizontal sub-surface water flows and surface run-off. This can produce a high volume of stormwater during rain and snowmelt events. The volume and energy of these can carry heavy loads of sediment and other pollutants into streams, and be very erosive of stream channels.

Roads convey stormwater as well as traffic. Rural roadside ditches drain and collect stormwater from both the impervious highway surfaces and adjacent land, and are usually designed to transport water to downstream outlets in natural streams. As a result, roadside drainage ditches are major conduits of stormwater discharge and pollution from the land to natural watercourses, providing an enhanced connection between land and water (Hoang and Porter, 2010). They are typically designed and maintained to convey highway drainage to outlets as expeditiously as possible, thus augmenting the energy, erosive force and capacity to convey pollutants of the flows they contain. Straight, steeply sloped and bare ditches scraped down to the substrate will carry far higher quantities of sediment than sinuous vegetated channels (Schneider,

2007). When such ditches discharge, the receiving stream or other water body may itself suffer bank or bed erosion as a result of that discharge.

There are thus potential conflicts of priority and trade-offs to be assessed in the management of ditches. As opposed to discharging drainage as quickly as possible, "stormwater management" seeks to reduce flooding, erosion and pollution impacts by retarding or retaining the water, providing for reduction in the energy of flow, recharge of groundwater and attenuation or deposition of potential contaminants (Schneider, 2007). Means by which the groundwater recharge may be achieved include permeable rather than impervious pavements, detention ponds and infiltration basins. It is also preferable to maintain ditches in a vegetated state, for example, through hydro seeding and mowing. Alternatively, they may be rock-lined to promote the ditch acting as an infiltration channel. In one study, a vegetated swale modified with a rock and soil check dam system designed to detain stormwater, reduced total phosphorus in captured run-off by as much as 54 percent and total suspended solids by 52 percent (Elfering and Biesboer, 2003).

A further problem for ditch management is that rural roadside ditches follow the road and therefore often pass through different political jurisdictions, each with varying management policies. For many ditches, the agency or individual with responsibility for managing the ditch may be unclear. For example, under New York Highway Law when a ditch falls within the road right-of-way, the highway manager responsible for the road is also responsible for the ditch. Conversely, if the ditch is not in the right-of-way, but lies in the private property immediately adjacent to the road, then the landowner may be responsible for maintaining it. In new residential subdivisions a homeowners' association established to manage affairs of the development may or may not assume the responsibility to maintain roads in the subdivision and associated drainage ditches. In practice it is possible that nobody takes responsibility for a ditch.

Placing management of rural roadside ditches within a watershed context recognizes their cumulative impact as conduits for land and highway drainage. Such cumulative impacts can be highly significant in causing the pollution of surface waters. To reduce such risk, the USC promotes ditch management that is coordinated and integrated by the responsible local and state government agencies. However, the legal and regulatory framework depends upon the scientific basis for this. The USC also provides training programmes, evaluates BMPs for ditch management and provides other forms of technical assistance to local municipalities in the Upper Susquehanna River Basin.

Thus institutional development parallels the practicalities of stormwater management. For example, within the USC two county stormwater sub-coalitions have formed (Chemung County Stormwater Coalition and Broome Tioga Stormwater Coalition). These sub-coalitions comprise leading water resource staff from county agencies, area colleges, engineering and architecture firms and other stakeholders working with communities and institutions affected by stormwater regulations. Such collaborations provide the capacity to develop innovative and cost-effective approaches that yield long-term results, rather than simply meeting minimum regulatory requirements.

### Wetland restoration

The USC has a wetland programme for which three major goals are identified, each specific to issues of concern in the Upper Susquehanna River Basin. These are to attenuate floods, enhance water quality and increase habitat and wildlife diversity. Wetlands, especially in the headwaters of a watershed, can desynchronize rainfall and run-off events through their vegetation and water-holding capabilities, thus reducing flood peaks, the energy of flows and downstream erosion. To benefit water quality, wetlands can retain sediment and nutrients during rainfall events and act as important nutrient and sediment sinks. Wetland complexes also provide unique habitats that can increase species diversity and habitat connectivity.

Under the USC programme a wetland team has been developed, including a wetland coordinator, two wetland biologists and several field technicians. Through various wetland grants and wetland mitigation projects the USC has been able to generate sufficient funding to purchase and employ operators for an array of wetland construction equipment, including a bulldozer, five excavators of various sizes, two tracked dump trucks, a hay mulcher, tracked compact loader and a building to house them. Thus the USC wetland programme can now offer a specialized "vertically and horizontally integrated" capability to complete any wetland project; combining expertise in wetland site identification and evaluation using GIS and other computer-based design tools, delineation, survey, design, construction and monitoring.

The team has constructed over 200 hectares of wetlands on both private and public lands using a diverse array of funding sources. Most significantly this USC activity has also spun off the creation of a partner not-for-profit organization, the Wetland Trust, which acts as the land steward for the USC when it has the strategic opportunity or need to achieve permanent land protection (the operation and funding of the Wetland Trust is discussed further below).

## Getting things done

### Ongoing challenges for planning and implementation

Under the President's Executive Order each Bay state was required to develop a Watershed Implementation Plan (WIP) to achieve and maintain the nutrient and sediment pollutant loads required by the Bay TMDL. The Chesapeake Bay TMDL WIPs should identify how the Bay jurisdictions are putting the measures in place needed to restore the Bay by 2025, and to achieve at least 60 percent of the necessary nitrogen, phosphorus and sediment reductions compared to 2009 levels by 2017 (USEPA, 2012). It is recognized that much of this work is already being implemented by the jurisdictions consistent with their existing commitments and restoration efforts (for example, New York State Department of Environmental Conservation published a Tributary Strategy for Chesapeake Bay Restoration in partnership with the USC in 2007).

The WIPs are intended by the USEPA to serve as "roadmaps" for how and when each state plans to meet its pollution allocations under the TMDL. To assist with WIP preparation the USEPA provided each state with detailed expectations for the plan, some technical and financial assistance and target pollutant loads. By the end of 2010 the USEPA concluded that the pollution controls identified in New York's draft WIP were sufficient to meet the TMDL, but the issue of whether full implementation can be adequately resourced remains unresolved. The estimates of ability to meet the state's TMDLs made by the USC and its partners were realistic but inevitably based on "best-guess" scenarios of required funding. Given the controversy and concerns about the technical feasibility and equity of the pollutant load reductions required by the TMDL allocated to New York State discussed above, timely provision of sufficient federal and state funding through to the target date of 2025 remains uncertain (NYSDEC, 2013)

Given funding, the keys to the success of the WIP in New York State (and of the Tributary Strategy that preceded it) are the availability of competent technical staff to assist local governments, farmers, other landowners and the general public in identifying and implementing appropriate management measures to enable the Bay's TMDL to be met, and the scientific support and understanding necessary to underpin this and give it credibility. The Upper Susquehanna Coalition demonstrates that county SWCDs are uniquely well-equipped to deliver the former and to develop, coordinate and employ the latter.

### USC governance, funding and capacity

Established in 1992 as an informal collaboration between county SWCDs, the USC transitioned in 2006 to a pure "conservation district coalition" using the Memorandum of Understanding described above. This MOU is based on New York and Pennsylvania state laws that allow districts to enter into multi-district agreements. The Tioga County Soil and Water Conservation District is designated in the MOU as the USC administrator, responsible for all contractual and other legal obligations.

The legal status of the USC and its institutional standing are important. New York State Soil and Water Conservation District Law Article II, Section 10, allows district directors to cooperate with one another to exercise the powers conferred by the SWCD law. Article II, Section 9(8) empowers directors of a soil and water conservation district to act as an agent for the United States, any of its agencies, New York State and any state agencies (the USC is also recognized as a bi-state coalition with respect to Pennsylvania). The conservation districts in the Upper Susquehanna region, including its Pennsylvania members, created the USC under these two provisions, and thus the USC appears to have the institutional standing and statutory mandate to act as an agent for federal and state agencies (and thus to receive funding directly from them). In practice however, the role of the USC as a partner with the state and the federal government is unevenly respected. For example, when federal funding becomes available for stormwater management, that funding is generally allocated by the USEPA as "state funds" to the designated state lead agency; in New York State that is the New York State Department of Environmental Conservation. The role of USC is then constrained because it lacks the institutional standing to directly receive that funding or to participate in the decision-making process for its allocation. The USC remains the most effective and best-informed agent of watershed management but its influence in the allocation of stormwater funding depends on decisions made by state officials.

Reputation and performance have overcome such limitations in the main to date, and federal and state sources provide the greater part of the funding for the USC. Most funds for planning and implementation are obtained competitively on behalf of the entire USC or for one or more of its counties. Operationally, the USC conducts multiple projects on varying watershed scales with local, regional, state, federal, academic and non-governmental partners. Over recent years the USC has obtained approximately US $3 million in funding, in addition to an equivalent amount generated through the federal Farm Bill for its Environmentally and Economically Sustainable Agriculture programme.

The technical capacities of the USC members encompass all aspects of non-point source pollution and watershed management. Staff members are especially well-trained and experienced in farm best management practices, stormwater management, wetland restoration, and stream corridor and floodplain management. As noted this "field" expertise is backed up by a strong scientific capability for modeling, planning, monitoring and evaluation. The USC members, as county organizations, are well-known locally, and earn and retain local respect and trust. This is crucial for their acceptance by farmers, community leaders and other stakeholders. This attribute of political acceptability of the SWCDs to local government leaders and their constituents is vital in fostering the local adoption of best management measures to meet the Bay TMDL. Overall these capacities, and the scope of the USC's operations in every county across the river basin, make it an indispensable partner to undertake the development and implementation of the state's Chesapeake Bay Watershed Implementation Plan under the auspices of the State Department of Environmental Conservation.

### The Wetland Trust

The Wetland Trust (Wetland Trust, 2013) is one of the first in the United States to develop an "in-lieu-fee" programme, and offers this to meet the needs for wetland impacts mitigation in the Upper Susquehanna. An in-lieu-fee programme is formally established by the US Army Corps of Engineers giving permission through an in-lieu-fee wetland instrument to either a non-profit organization such as the Wetland Trust or a governmental entity. Under this instrument the Wetland Trust collects funds from any entity required to effect compensatory mitigation because they impacted wetlands as described under the federal Clean Water Act, Section 404 or other wetland regulatory law. The Trust applies these funds, aggregated from multiple such permittees, to create one or a number of wetland sites to satisfy the mitigation required of the permittees. In-lieu-fee mitigation is generally mitigation that is provided after the permitted impacts or wetland losses have occurred and thus takes place when a permittee provides funds to the Trust rather than completing wetland mitigation themselves, or alternatively purchasing credits from an approved wetland mitigation bank.

The Wetland Trust is providing high-quality wetland projects in the Upper Susquehanna Watershed, and continues to increase its list of high-priority potential sites as it implements the in-lieu-fee programme. Acquisition of land ownership for the wetlands to be restored or created is required by the Trust as this guarantees long-term protection. The Trust also benefits from its close association with the USC, as the USC can provide scientific and technical support for site identification and planning, and can be hired to conduct the restoration works on the properties that the Trust acquires.

## Outcomes

It is impossible to disaggregate and attribute USC contributions alone in the meeting of water quality goals and sustainable management of the water resources of the Upper Susquehanna, but as noted above both the Susquehanna and its Chemung River tributary have in the main good water quality. Since its creation the USC has played a major part in sustaining this and has had a cumulative impact in fostering improved water quantity and quality management in the Upper Susquehanna River Basin. It has worked with municipalities, landowners and other stakeholders, and its three primary component programmes have demonstrated the ability to promote acceptance of land and water management practices that effectively protect water and related natural resources. Box 3.1 provides a case study of success achieved by measures under the Environmentally and Economically Sustainable Agriculture programme. A wide range of information sheets and other technical and outreach resources for stakeholders, USC members and partners to use are also made available from the USC website and its members' county offices.

## Box 3.1 Farm BMPs upstream put lake on the path to recovery

Stephen Foster Lake is in Mt. Pisgah State Park in the northern mountain region of Bradford County, Pennsylvania. A renowned fishery, the lake encompasses 28 hectares, has an average depth of 3.2 meters and receives 150,000 recreational visitors each year. Over half of the lake's 2,660-hectare watershed is used for agriculture; the remainder being predominantly forested. Over time, Mill Creek, the feeder stream, deposited excess sediment and nutrient run-off in the lake, creating anoxic conditions. Large, unsightly algal blooms reduced the available oxygen for aquatic organisms, including the fish species that attracted visitors. In 1996 Pennsylvania added the lake to the state's list of impaired waters for nutrient and sediment impacts from agriculture. The lake will not meet its designated recreational uses until algal blooms no longer manifest and the Trophic State Index (TSI) declines.

Bradford County Conservation District (BCCD) and its restoration partners in the farming community worked together to address these issues. By 2004, 11 of the 13 farms in the watershed upstream had fully implemented agricultural BMPs. These included over 14 kilometers of stream fencing and alternative water supply systems to keep livestock from waterways, improved crossings to swiftly move cattle across streams without stream bank grazing and erosion, improved manure treatment and storage on each farm, planted riparian buffers and almost a kilometer of restored stream channel.

These BMPs dramatically reduced the amount of sediment and nutrients delivered to the lake. Computer models calculated that the BMPs reduced phosphorus and sediment run-off loads by 52 and 59 percent respectively, exceeding the TMDL-specified reductions. By 2005, biological monitoring showed improvements in biological conditions in the stream, while lake water quality data reflected slight decreases in the levels of total phosphorus, total suspended solids and TSI value. More substantial lake water quality improvements were still needed but were expected to emerge slowly because of the large amounts of legacy sediment that release phosphorus during seasonal periods of low dissolved oxygen. Additional in-lake treatments are being researched and implemented to treat the phosphorus-laden lake sediment, now that pollutant inputs from the watershed have been controlled.

The farm BMP implementation was made possible by US $274,000 of Clean Water Act Section 319 funding and technical assistance from the USEPA. Pennsylvania Department of Environmental Protection, BCCD, US Department of Agriculture, Chesapeake Bay Foundation, Pennsylvania Department of Agriculture, Natural Resources Conservation Service and landowners together provided further funding for all components of this project totaling US $1.2 million.

*Source: M. Lovegreen, Bradford County Conservation District, Pennsylvania*

With the USC's guidance available when needed, most farmers in the upper watershed now participate in the New York State AEM programme and thus have instituted nutrient management plans and adopted best management practices. The USC's Riparian Buffer Initiative has also been highly effective in protecting stream corridors. However, estuaries are the foot of the hydrological "hill" and act as repositories for accumulated nutrients and sediment carried to them by their tributaries. As land uses increase and population grows in its watershed so the condition of the Chesapeake Bay deteriorates. Restoration now depends upon achieving its TMDL, but despite the best efforts of the USC the public response to date raises concerns. In numerous meetings arranged by the USC with agencies and stakeholders at all levels, non-acceptance of the load targets set by the Chesapeake Bay Program has been expressed as the loads were lowered to levels perceived to be unobtainable, and inequitable in relation to other states. Most importantly, this includes representatives of the farming community. Even assuming adequate funds are made available, a caution has been expressed by the latter that the prospective load targets may be impracticable as there is a threshold beyond which farmers simply cannot achieve or sustain additional water quality improvements.

For stream restoration local delivery teams of SWCD personnel are being trained by a natural stream design specialist to collect data and analyse stream reaches, perform "triage" (i.e. reject those with no hope of help) and design and implement rehabilitation projects. Training of local community volunteers is also underway to enable them to conduct simple stream assessments as an outreach and public education tool. Many kilometers of streams have now been rehabilitated by this programme and significant reductions in stream bank and stream bed erosion throughout the watershed are observable.

For unpaved road and road ditch improvement a GIS tool has been developed from a successful dirt and gravel road programme by Pennsylvania State University. This is being used to inventory watersheds to locate severely eroded road banks and ditches and to estimate the sediment loads from these sites. Data collected assists in road ditch restoration projects and in providing training for local highway personnel.

The USC wetland programme has gained a high regard in its efforts to reduce flooding, water quality degradation and increase habitat and wildlife diversity. The USC is continuing to develop this programme as a "showcase" of integrated natural resource restoration approaches that can address all aspects of wetland issues. The recognition and standing achieved by the USC wetland programme provided the foundation for the creation of the Wetland Trust, while the Trust was a necessary niche activity for the USC to enable holding of fees for mitigation.

## Concluding reflections

### *Strengths and lessons of success*

Key lessons from the Chesapeake Bay Program and USC experience can be listed as follows, and are discussed further below.

*   Legitimacy and institutionalization of governance arrangements;
*   Cooperative partnerships within and between different levels of government;
*   Sound science and technical knowledge;
*   Effective local technical capacity through technical providers and enablers;
*   Transparent governance;
*   Accountability for stewardship and performance;
*   Public participation opportunities.

A key success has been establishment of the reputation, standing, legitimacy and institutionalization of the USC and its governance. It has become the principal watershed-wide organization operating in the headwaters of the Susquehanna River, and a trusted delivery partner for the New York State Department of Environmental Conservation and other state agencies.

The USC also demonstrates the importance and effectiveness of cooperative partnerships within and between different levels of government and non-government organizations; for which again its status and standing matters. The cooperative partnerships fostered by the USC are excellent. The member SWCDs of the USC work closely with units of local government, state-level agencies such as the New York State Departments of Agriculture and Markets, Environmental Conservation and Transportation, and at the federal level with the US Department of Agriculture and the USEPA, who are also primary sources of funding for USC projects.

Involvement of the scientific community in the Chesapeake Bay Watershed Program is one of its critical components. The importance of programme design and decision-making based on the best available science and USC partnerships with regional universities has been highlighted above. Models are essential tools for understanding watershed processes, particularly for assessing the sources and loading of pollutants by sub-catchment or other sub-division. However, the experience of the Chesapeake Bay TMDL allocation for New York State well-illustrates that models must by credible, trusted and accepted by all who use them and are impacted by the decisions based on their use.

An essential requirement for implementation of a watershed improvement programme is effective technical capacity delivered by technical providers and enablers. The core goals of SWCDs are to support local efforts to improve air and water quality, reduce soil erosion and non-point water pollution sources. This is achieved through conservation planning and the design and implementation of innovative best management practices. SWCDs also have a lead technical role for local flood management and mitigation. SWCD personnel generally have a wide technical background and can act as enablers for this range of projects and programmes. Good communication and personal skills are required to work with landowners to develop and implement conservation plans and practices, and also develop strong collaborative relationships with partner agencies.

SWCDs depend upon the trust, acceptance and respect of the constituencies with which they work. Their offices are open to the public and are friendly and accessible. Farmers, small business owners and householders can walk in to obtain free advice on regulation and management practices to meet that regulation. Such advice is provided from an independent and non-regulatory stance. SWCDs work with the individual to jointly solve a problem through voluntary action and regulatory compliance, and would only turn to a regulatory approach in the case of severe serial offenders non-responsive to all engagement efforts. They maintain strong outreach efforts through their websites and newsletters, and their underlying aim is to give the local communities a sense of ownership in the SWCDs and their operations. State and federal agencies are also generally open and freely convey information about their decisions and the basis for these. Thus environmental governance for the Upper Susquehanna River Basin, at both local and higher levels, can be considered "transparent" and open to public scrutiny.

There is full accountability for stewardship and performance at all levels. USC members comprising the SWCDs of 19 counties are all answerable to their district board of directors. They are also accountable to county legislatures, which provide significant funding to the SWCDs as county departments. The SWCDs under the New York State Soil and Water Conservation District Law are also answerable to the State Soil and Water Conservation Committee within the State Department of Agriculture and Markets. The state agencies involved in the Chesapeake Bay Program are accountable to their respective state legislatives. The USEPA is answerable to the Executive Branch (the White House) and to the US Congress. All these elected bodies provide ongoing scrutiny. A readiness of any disgruntled organizations and citizens to sue also means that the agencies are all potentially accountable in courts. For example, the citizen actions against the USEPA to enforce the application of the Clean Water Act, Section 303, led directly to the current TMDL programme.

Successful watershed management actions and their accountability both depend on communication, engagement and understanding and genuinely open partnerships between the multiple constituencies concerned. The decisions of farmers, other landowners, local government leaders and non-government organizations all contribute to the objectives of the watershed management programme. Hence considerable efforts are sustained at federal, state and local levels, such as the USC, to inform and involve stakeholders and the wider public. However, public support will depend on the public's perception that watershed measures are both economically efficient and equitable. The USC strives to meet these criteria at a local level in its programme delivery, but for the taxpayers of New York State the simply proportional allocation of TMDL reductions to Bay states fails on both counts as described above.

### Continuing challenges

There are two major continuing challenges for the USC: the challenge of sustaining the capacities of its staff and renewing leadership, and the challenge of maintaining cooperative partnerships and an institutionally recognized and supported watershed role for the coalition.

Since its formation in 1992 the USC has benefited from effective and innovative coordination and leadership, both at the level of the coalition itself on the part of the coalition coordinator and at member level in the county SWCDs that comprise the coalition. Partnership working with a diverse range of government and non-government organizations, bidding for funds, linking with the scientific community, public participation and institutional development of the USC and the Wetlands Trust all require skills in leadership and as an effective intermediary. The USC is faced with the challenge of sustaining this capacity as its leaders retire or take other positions. Conscious preparations are being made for these transitions by nurturing younger staff to assume greater responsibilities. The quality of an organization is inseparable from the quality of its personnel. The success and sustainability of an organization ultimately depends upon the strengths of the individuals within it, and this cannot be taken for granted. An important corollary to leadership is that it has been completely supported by funds generated by the coordinator and senior USC staff; no county district funds support the USC efforts as they are used for county projects.

Regarding the institutional challenge, leaders and senior staff in the agencies with which the USC partners also retire or move to other positions. It is naive to assume the memories of past watershed accomplishments always survive these inevitable departures. This reality reinforces the necessity for the USC to gain stronger institutional standing through legislation that explicitly recognizes and authorizes the watershed role and duties of the Upper Susquehanna Coalition.

## Notes

1   Detailed watershed maps can be viewed from the websites of the Chesapeake Bay Program (http://www.chesapeakebay.net/discover/baywatershed) and Upper Susquehanna Coalition (http://www.u-s-c.org/html/Resources.htm#maps).
2   This follows a report in 1966 (National Association of Soil and Water Conservation Districts, 1966) by the SWCDs' national organization (later the National Association of Conservation Districts) which urged districts to be inclusive and to be the natural resources representative not only of agriculture but also of business, industry, recreation and community interests (Helms, 1992).

# References

Bennett, H. H. (1943) 'Adjustment of agriculture to its environment', *Annals of the Association of American Geographers*, vol. XXXIII, pp. 163–198.

Davison, S. G., Merwin, J. G., Capper, J., Power, G. and Shivers, F. R. (1997) *Chesapeake Waters: Four Centuries of Controversy, Concern, and Legislation*, Tidewater Publishers, Centreville, MD.

Elfering, J. and Biesboer, D. D. (2003) *Improving the Design of Roadside Ditches to Decrease Transportation Related Surface Water Pollution*, Minnesota Department of Transportation, Research Services Section, St Paul, MN.

Helms, D. (1992) 'Getting to the roots' in *People Protecting Their Land: Proceedings Volume 1*, 7th ISCO Conference Sydney, International Soil Conservation Organization, Sydney, Australia, pp. 299–301.

Hoang, Y. and Porter, K. S. (2010) 'Stormwater management in the rural New York headwater areas of the Chesapeake Bay Watershed', *The Journal of Water Law*, vol. 21, no. 1, pp. 6–18.

Horan, E. (2005) *Memoranda of Understanding: Delegating Authority to Substate Governmental Entities*, Water Law Clinic, Cornell Law School.

Horton, T. (2003) *Turning the Tide: Saving the Chesapeake Bay*, Island Press, Washington, DC.

Langland, M. J., Edwards, R. E., Sprague, L. A. and Yochum, S. E. (2001) *Summary of Trends and Status Analysis for Flow, Nutrients and Sediments at Selected Nontidal Sites, Chesapeake Bay Basin, 1985–99*, U.S. Geological Survey Open File Report 01-73, New Cumberland, PA.

National Association of Soil and Water Conservation Districts (1966) *The Future of Districts: Strengthening Local Self-Government in Conservation and Resources Development*, Report of the Special Committee on District Outlook, National Association of Soil and Water Conservation Districts, League City, Texas.

NYSDEC (2007) *The 2004 Chemung River Basin Waterbody Inventory and Priority Waterbodies List*, New York State Department of Environmental Conservation, Albany, NY.

NYSDEC (2009) *The 2009 Susquehanna River Basin Waterbody Inventory and Priority Waterbodies List*, New York State Department of Environmental Conservation.

NYSDEC (2013) *Final Phase II Watershed Implementation Plan for New York Susquehanna and Chemung River Basins and Chesapeake Bay Total Maximum Daily Load*, New York State Department of Environmental Conservation, www.dec.ny.gov/docs/water_pdf/finalphaseiiwip.pdf, accessed September 3, 2013.

Porter, K. S. (2004) 'Does the CAP fit? Rectifying Eutrophication in the Chesapeake Bay', *The Journal of Water Law*, vol. 15, no. 5, pp. 188–192.

Porter, K. S. (2010) 'The Chesapeake Bay Watershed: A Legal Battlefield', *The Journal of Water Law*, vol. 21, no. 4, pp. 114–122.

Schneider, R. (2007) *Roadside Ditch Management to Reduce Stormwater Runoff, and Mitigate Floods and Droughts*, presentation to a meeting of the Delaware River Basin Commission, December 12, 2007, Trenton, NJ.

Smith, D. E., Leffler, M. and Mackiernan, G. B. (1992) *Oxygen Dynamics in the Chesapeake Bay: A Synthesis of Recent Research*, Maryland Sea Grant College, Maryland University, College Park, MD.

UMCES (2010) *Ecocheck, Chesapeake Bay – Overview: 2010*, www.eco-check.org/reportcard/chesapeake/2010/overview/, accessed August 15, 2013.

UMCES (2012) *Chesapeake Bay Health Improves to a C in 2012*, www.umces.edu/story/2013/jul/03/chesapeake-bay-health-improves-c-2012, accessed September 2, 2013.

UMCES (2013) Chesapeake Report Card 2013, www.ian.umces.edu/ecocheck/report-cards/chesapeake-bay/2013/, accessed January 3, 2015.

USC (2013) *The Upper Susquehanna Coalition*, www.u-s-c.org/html/index.htm, accessed September 3, 2013.

USEPA (2008) *Bay Barometer: A Health and Restoration Assessment of the Chesapeake Bay and Watershed in 2008*, Chesapeake Bay Program.

USEPA (2010) *Fact Sheet Chesapeake Bay Total Maximum Daily Load (TMDL)*, Chesapeake Bay Program.

USEPA (2012) *Chesapeake Bay TMDL*, www.epa.gov/reg3wapd/tmdl/ChesapeakeBay/tmdlexec.html, accessed August 15, 2013.

Wetland Trust (2013) *The Wetland Trust*, www.thewetlandtrust.org/, accessed September 3, 2013.

# 4 New York City Watershed Protection Program

## A national paradigm?

*Keith Porter and Laurence Smith*

## Introduction

Safety of public water supplies has long been taken for granted in the United States. However, in recent years newly recognized threats have compelled a critical re-examination of how to best protect water supplies. Disinfection and filtration methods relied upon since the early 1900s now seem insufficient. Apart from a multitude of chemicals and their residuals that are by-products of our modern economy there are other risks posed by microbes such as the protozoan parasites: *Cryptosporidium parvum* and *Giardia lamblia*. The possible insufficiency of conventional water treatment alone to protect the wholesomeness of drinking water has reawakened interest in protecting drinking water supplies at their source as a first barrier of defense against water-borne contamination. The source of water supplies is the catchment, or watershed in US parlance.

New York City is a thirsty city. Since its first settlement in the early seventeenth century sufficient water of wholesome quality has been a driving concern. The failure of the Dutch governor, Peter Stuyvesant, to ensure a water supply to Fort Amsterdam was instrumental in his surrender to the British in August 1664. In the first decades of the nineteenth century population growth and consequent stresses on local wells compelled New York City authorities to seek new and more reliable sources of drinking water. This led to its first "extra-territorial" development of new reservoirs in the Croton River system north of the City. Over the next 100 years New York City continued to seek and develop new sources of water further and further from the city itself, reaching as far as the Delaware River system in the 1950s. The result today is an impressive water supply system of over 20 reservoirs and managed lakes within a total catchment area of over 5,000 square kilometers.[1] This system has subsequently ensured the city has a high-quality source of drinking water, and led to an innovative programme of watershed management to protect the city's water supplies at their source. As discussed further below this protection at source is especially critical for public health and legal reasons, not least because the city's water supply is not filtered before delivery to water consumers. In recent years, the question of sustaining the high quality of this water supply without the use of filters has become a major and contentious political issue with very high stakes.

Protecting drinking waters at their source, or source water protection, is an objective of the federal Safe Drinking Water Act (42 U.S.C.A. Sections 300j-13, 300j-14. 1996). A nationally foremost demonstration of source water protection is the New York City "watershed." About 10 percent of the watershed area used for the city's water supply is immediately north of the city and consists of urbanized and suburban land uses. This is the Croton part of the watershed system. The other 90 percent is on the western side of the Hudson River and includes the Catskill Mountains, a largely forested area famed for its natural beauty. This watershed region has a population of about 72,000, giving a very low population density of about 15 persons per square kilometer. The land use patterns are mainly small hamlets, villages, and family farms, all surrounded by large areas of open space or forest.

Farming and other human activities create potential pollution. To securely protect the water supplied by such an inhabited watershed it is essential that farmers and other landowners, businesses, community leaders, and residents willingly manage the sources of potential pollution they control. Accordingly management of non-point sources in particular is a matter of local management. This has implications for the design of any source water protection programme and for the governance arrangements necessary to deliver and sustain it. Effective watershed management requires sharing and coordination of authorities and roles between different levels of government. A balance is also required between regulations and voluntarily adopted best management practices. As in the example of the New York City Watershed, such balancing of responsibilities may range in scale from involvement of federal and state authorities in relation to inter-river basin transfers of water, to respecting the private property rights of individual farmers and homeowners.

Interbasin transfers provide a major part of New York City's water supply. The waters in Delaware County on the western side of the Catskills are the headwaters of the Delaware River. In order for New York City to obtain water from this river system it required interstate approval. This was opposed by New Jersey. The Delaware River is the major river in New Jersey and that state objected to the proposed transfer of a large amount of water from its basin to New York City. In 1931, the US Supreme Court famously decided the case, *State of New Jersey v. State of New York*, (283 U.S. 336, 1931), one of the most notable cases of interstate stream litigation. The state of New Jersey sued to enjoin New York City and New York State from transferring waters from the Delaware River to New York City. In its decision, the Supreme Court decreed that New York City could divert up to 440 million gallons of water per day (mgd) from the Delaware River. On June 7, 1954, the Court increased this amount to 800 mgd. These decisions allowed the construction of a dam on the Delaware River in Delaware County to create the Cannonsville Reservoir. This reservoir is the third largest of the 20 reservoirs in the New York City water supply system with a volume of nearly 100 billion gallons.

In delivering the decision of the Supreme Court in 1931, Mr. Justice Holmes stated: "A river is more than an amenity, it is a treasure. It offers a necessity of life that must be rationed among those who have power over it." Current governance of the New York City Watershed inverts this statement to "how are the powers of the various jurisdictions governing the New York City Watershed to be rationed over the necessity of life that its reservoirs now supply?" This chapter describes that sharing of powers in what has been heralded as a prototype watershed programme of the utmost importance to all water supply managers (Committee to Review the New York City Watershed Management Strategy, 2000).

## Context and drivers

### *The Safe Drinking Water Act*

In 1986 the 99th Congress enacted substantial amendments to the Safe Drinking Water Act (SDWA). Congress was concerned with newly identified chemical and microbial threats of contamination to drinking water supplies and the adequacy of federal standards in meeting those threats (Kelly, 1986). Accordingly, a main purpose of the SDWA amendments was to strengthen drinking water standards and enforcement, and to provide greater groundwater protection. It was the intent of Congress to require water supplies generally to provide filtration in their water treatment plants. Crucially, in exceptional cases such as the New York City water supply system, Congress provided for filtration avoidance. The amendments therefore also required the administrator of the US Environmental Protection Agency (USEPA) to specify "criteria under which filtration ... is required as a treatment technique for public water systems supplied by surface water sources. In promulgating such rules, the Administrator shall consider ... protection afforded by watershed management" (Safe Drinking Water Act 42 U.S.C.A. Section 300g-1(b) (7)(C)(i)).

Such provisions would suggest the Act specifically provided for the protection of drinking water supplies at their source. However, in the 1986 amendments this specific provision was only for sole source aquifers and wellhead areas. Surface waters were not accorded such explicit protection until measures to promote the protection of water supplies at their source were adopted in 1996. These new 1996 provisions included a source water quality assessment programme and a source water petition programme. The latter has the aim of "facilitating the local development of voluntary, incentive-based partnerships among owners and operators of community water systems, governments, and other persons in source water areas" (Section 300j-14(a)(2)(A)). Since the 1986 amendments, the quality and protection of source waters has thus assumed critical importance for unfiltered water supply systems such as that of New York City.

*Filtration avoidance*

In a critical clause the Safe Drinking Water Act Amendments of 1986 mandated:

> Not later than 18 months after the enactment of the Safe Drinking Water Act Amendments of 1986, the Administrator shall propose and promulgate national primary drinking water regulations specifying criteria under which filtration (including coagulation and sedimentation as appropriate) is required as a treatment technique for public water supply systems supplied by surface water sources. In promulgating such rules, the administrator shall consider the quality of source waters, protection afforded by watershed management, treatment practices (such as disinfection and length of water storage) and other factors relevant to health.
>
> (Section 1412 (b)(7)(C)(i))

Since New York City does not filter its water supplies it sought to comply with the criteria after their promulgation and applied for "filtration avoidance." The basis for meeting the criteria was the high quality of the water obtained from the Catskill–Delaware watershed system without filtration. As Dombeck (2003, p. 125) states: "New York City has some of the best water in the world, and the reason is trees." A high proportion of the watershed is forested and this together with the low population density accounts for the high quality of water the watershed yields. If compelled to construct filters, the City estimated the capital costs in 1990 to be US $8.0 billion (Galusha, 1999). Watershed protection seemed to offer a substantially less expensive alternative. Accordingly, in 1990, the City proposed new Watershed Rules and Regulations to control potential sources of pollution in the watershed. These "command and control" proposed regulations provoked furious anger among farmers and watershed communities. The assumption of police powers through the proposed regulations was seen as threatening livelihoods and the way of life in the watershed.

    Others, including engineers in the New York City Department of Environmental Protection, shared doubts about the alternative of source water protection and argued strenuously that the City should construct filters to ensure the security of the City's water supplies (Kennedy, 1997). In fact, convictions about the necessity of providing filters for water supplied from an area with farming dates back to the early use of filters. A water company supplying water, without filtration and obtained from unfenced reservoirs fed by streams receiving drainage from farmyards, was held not to be supplying pure and wholesome water as required by the Waterworks Clauses Act, 1847 (*Attorney-General v. Rhymney and Aber Valley Gas and Water Company* (1907), 71 J. P. 435).

    For New York City's water supply system there was also influential advocacy for filtration at the national level. An expert panel convened by the USEPA expressed emphatic endorsement of filtration as the primary way to protect the City's water supply (Okun *et al.*, 1993). Five members of the panel were sufficiently opposed to "filtration avoidance" to publish an article in the *Journal of the American Water Works Association* after the panel delivered its report to the EPA (Okun *et al.*, 1997). The five panel members stated: "Without filtration, given the degree of development that already exists on New York City's watersheds and the City's limited capacity to restrict further development, the people of the city, particularly those who are immunocompromised, are potentially at risk" (Okun *et al.*, 1997, p. 63). Eileen Pannetier, a water quality specialist with Comprehensive Environmental Inc, Dedham, Mass, was the sixth member of the panel who declined to be a co-author of the article. She regarded it as inappropriate for panel members to comment on events after the panel was dissolved. The USEPA insisted on filters in response to an analogous application for avoidance submitted by the Massachusetts Water Resources Authority on behalf of the Boston water supply system (Kavanaugh, 1999). However, New York City's case for filtration avoidance was well argued in an accompanying article written in response to the objections of the five panel members to filtration avoidance (Ashendorff *et al.*, 1997). An underlying assumption in the arguments is that comprehensive watershed management is an alternative to filtration. Conversely, if filters were constructed they would be a primary barrier relied upon, and not protection of the raw waters at their source. This was expressed by the statement approved in 1989 by the Council of the New York Academy of Medicine:

> The New York Academy of Medicine cautions against relying solely on filtration to protect the public. Many examples exist of waterborne disease epidemics caused by failures of filtration systems. The advent of filtration also will serve as a disincentive to source water protection, causing increasing pollution loadings that in turn pose greater health risks and necessitate expensive modifications to filtration systems.
>
> (Committee on Public Health, 1989, p. 900)

In 1989, the USEPA promulgated the Surface Water Treatment Rule (SWTR) regarding filtration avoidance. Criteria specified by the SWTR to avoid filtration included limits for fecal coliform and turbidity, and disinfection and monitoring requirements. In addition, a principal criterion was adequate watershed protection (40 C.F.R. Section 141.71). As specified under the regulations: "The public water system must maintain a watershed control program which minimizes the potential for contamination by *Giardia lamblia* cysts and viruses in the source water" (40 C.F.R. § 141.71 (b)(2)). In addition, the watershed control programme must:

1 characterize the watershed hydrology and land ownership;
2 identify watershed characteristics and activities which may have an adverse effect on source water quality; and
3 monitor the occurrence of activities, which may have an adverse effect on source water quality.

The regulations further require: "The public water system must demonstrate through ownership and/or written agreements with landowners within the watershed that it can control all human activities which may have an adverse impact on the microbiological quality of the source water."

## Getting started

### *The New York City Watershed Memorandum of Agreement (MOA)*

Following the rejection of its draft regulations in 1990 by watershed communities, New York City continued seeking to meet the criteria for filtration avoidance through negotiations with watershed communities, regulatory agencies, and other concerned parties, over an extended period lasting six years. In early 1997, the governor of the State of New York, the mayor of the City of New York, the Coalition of Watershed Towns, 12 villages in the watershed, and a consortium of environmental groups, signed the New York City Watershed Memorandum of Agreement (MOA). This unprecedented agreement required New York City to invest about US $1.5 billion in upgrading wastewater treatment, stormwater management, and environmentally benign development. In addition, the MOA required the City to purchase land in the watershed, forcing it "to assume the role of country squire" in becoming one of the largest landowners in upstate New York (Hu, 2004). For the watershed communities a critical component of the MOA was its Article V. This set out obligations and mechanisms for payment of funds necessary to maintain and enhance both the water quality and the economic and social character of communities in the watershed. It was assumed that the funds would be provided by New York City. In other words, the consumers of the water supplied would meet the costs of the watershed protection programme out of their water rates. Rather than the polluter, the beneficiary pays. The programmes provided through Article V are collectively termed the watershed protection and partnership programmes. The MOA recognized that a working partnership between the City and the communities required that an independent, locally based and administered, not-for-profit corporation needed to be established in the Catskill–Delaware Watershed, To this purpose, the Catskill Watershed Corporation (CWC) was formed to administer and manage the following watershed protection and partnership programmes:

- Stormwater Retrofits;
- Sand/Salt Storage Facilities;
- Septic Remediation Program;
- Stormwater Fund;
- Public Education;
- Economic Development Study;
- Catskill Fund for the Future;

- Tax Consulting Fund;
- Land Acquisition Consultation.

The CWC is responsible for decisions regarding the funding and implementation of these partnership programmes. Its board of directors comprises 15 members, 12 of which represent locally elected officials of communities in the watershed. The governor of New York State appoints two members (one of whom is an environmental representative), and one member is appointed by the mayor of New York City. The CWC has also subsequently played a vital economic and educational role in funding and supporting communities and activities in the watershed.

In 1997 the USEPA announced a "filtration avoidance determination" (FAD) for the Catskill and Delaware region of the watershed. In doing so, the "EPA had to 'rewrite the book' to provide a filtration avoidance determination to New York City" (Rodenhausen, 2001, p. 471). Simultaneously, the New York State Department of Health approved the New York City Watershed Rules and Regulations as New York State regulations under Public Health Law Section 1100 at 10 NYCRR Part 128. A particular focus of these new regulations, by which the watershed would be governed, was the group of reservoirs stressed in terms of their water quality by excess nutrients and algal growth. The most significant of these reservoirs given its size was the Cannonsville Reservoir in Delaware County.

### Managing phosphorus

Water supplied by New York City to its nine million customers meets all drinking water standards, and "New York City's water supply is world renowned for purity and excellent taste" (Committee on Public Health, 1989, p. 898). However, historically there have been high levels of phosphorus in the Cannonsville Reservoir in the Catskill–Delaware system. Eutrophication and algal growth caused by high levels of the nutrient may lead to odors, tastes, and colors in the water supply. A particularly important threat occurs when chlorine reacts with organic material such as algal cells and detritus in the water to form disinfection by-products (e.g. trihalomethanes), which are a carcinogenic risk.

Under the MOA all reservoirs in the New York City Watershed must meet New York State (NYS) Water Quality Standards and Guidance Values. The MOA specifically states the "New York State guidance value for phosphorus will be used for Total Maximum Daily Load (TMDL) development for the reservoirs." New York State has a guidance value of 20 microgrammes per litre ($\mu$g/L) for phosphorus. The watershed regulations designate the drainage basins of the enriched reservoirs as "phosphorus restricted"; in other words a phosphorus-restricted drainage basin is one in which the phosphorus present produces concentrations exceeding the guidance value. In a basin that has been designated as "restricted" the regulations do not permit new wastewater treatment plants or expansions of existing wastewater treatment plants that discharge to surface waters; clearly a constraint to new housing or other commercial development. However, the regulations allow for variance in the discharge of existing plants if any increase in phosphorus is offset by a twofold reduction in phosphorus loading from another source elsewhere in the basin. The regulations also provided for a pilot programme under which new or expanded wastewater treatment might be permitted if phosphorus were reduced by three times the proposed discharge elsewhere in the basin (10 NYCRR Section 128-8.3(a)(1)). As a condition of this pilot programme the regulations required the preparation of a County Comprehensive Strategy in support of the request for the variance to be allowed.

### The critical case of the Cannonsville Reservoir and local initiative

The Cannonsville Reservoir, with a drainage basin of almost 118,000 hectares, has the largest catchment in the New York City Watershed. About 200 kilometers north-west of New York City, it was created by damming the west branch of the Delaware River. The reservoir holds almost 380 million cubic meters of water at full capacity, and daily supplies about 322,000 cubic meters or about 7 percent of the consumption of New York City. Its watershed has had a population for a century or more with the low density of about 17 persons per square kilometer, a slightly higher density than the Catskills as a whole. Forest, and abandoned farmland reverting to forest, is the predominant land use. The watershed consists of approximately 70 percent forest, 25 percent farmland and 5 percent urban. It has an attractive hilly terrain with narrow river valleys. Although the catchment has a low density of land uses, there are many family-owned dairy farms and the water in the Cannonsville Reservoir has been enriched by phosphorus. When the MOA was signed the reservoir's water exceeded the guidance value of 20 µg/L for phosphorus and accordingly its drainage basin was designated as "phosphorus-restricted."

This designation was an especially serious economic constraint for Delaware County, much of whose area is accounted for by the Cannonsville Reservoir's watershed. This county is one of the poorest in New York State and its business community was especially concerned about the economic consequences of the phosphorus restriction. Furthermore, Delaware County has a highly conservative self-reliant culture. Leaders in the county, along with other upstate rural communities, traditionally distrust government. This is more than sentiment as New York is recognized as a "home-rule" state under the New York State Constitution, Article IX, Local Governments. "Home rule" can be defined as a default presumption for local control over matters of local concern and local government leaders in Delaware County keenly defend such rights. However, the importance of "home rule" status does not need to be over-emphasized. As noted above, for all jurisdictions watershed management to control point and non-point source pollution will depend on the decisions of landowners, business persons, local government, and residents. These are local matters and responsibilities, usually subject to local regulations and authorities.

*Delaware County requests assistance*

In 1998, the Delaware County Board of Supervisors requested the New York State Water Resources Institute (WRI) at Cornell University to assist in creating a comprehensive strategy for the county as prescribed by the New York City Watershed Rules and Regulations Section 18-83. The chairman of the board, Ray Christensen, specified that the strategy should have sound scientific credentials to ensure its credibility and acceptance. He also insisted that the strategy should be developed with institutional partnerships at local, state, and federal levels. The WRI met the first condition by creating an inter-disciplinary scientific team. To satisfy the institutional condition, the Delaware County Board accepted the recommendation to establish two guiding working groups: an interagency policy committee, called the Delaware County Phosphorus Study Committee, and a scientific support group. All watershed partners at local, state and federal levels were represented on both groups (Delaware County Department of Watershed Affairs, 2002).

In 1999, the WRI prepared a draft strategy for the county termed the Delaware County Action Plan. The board of supervisors approved this strategy in September 1999. State and federal partners to the MOA subsequently endorsed the strategy. To implement the strategy, the board of supervisors of Delaware County established a new Delaware County Department of Watershed Affairs.

## Approach and tools

### *The Delaware County Action Plan through "technical providers"*

The Delaware County Action Plan (DCAP) fulfills MOA requirements for a Delaware County Comprehensive Strategy as required under the New York City Watershed Rules and Regulations. Its goal is to assist the county's residents, farmers, businesses, and communities in meeting water quality objectives while retaining economic vitality. The DCAP is a locally led instrument designed to act upon and build local capacity. Its primary agencies are county departments that work cooperatively with the watershed partners, especially the CWC and New York City, through coordination provided by the Delaware County Department of Watershed Affairs. Staff of the county agencies, especially the Departments of Economic Development, Planning, Public Works, and Watershed Affairs, and the Delaware County Soil and Water Conservation District, are key in providing technical assistance at the local level. These "technical providers" are respected and trusted locally as members of the local community. This is critical for their acceptance. Cornell University provided scientific and technical support through the WRI. Although established as a response to the phosphorus restriction the watershed partners insist that the DCAP is a comprehensive source water protection programme and addresses all potential contaminants.

The DCAP conducts planning, pilot projects, demonstrations, and management initiatives based on sound scientific credentials. Given the phosphorus restriction, the initial focus of the DCAP was on the Cannonsville Reservoir basin. Phosphorus loads from point and non-point sources in the watershed have been measured since 1975, and intensively monitored from 1991 (Longabucco and Rafferty, 1998). Annual loads have varied from about 20,000 kilograms to a peak of 174,000 kilograms. At the time of the phosphorus restriction, the average annual load was about 50,000 kilograms. This annual load of phosphorus was sufficient to maintain concentrations of phosphorus in the reservoir at 20 μg/L or above. Calculations suggested the overall average annual loading of phosphorus to be sought should be 40,000 kilograms or less. Therefore the target reduction sought was 10,000 kilograms per year or about 20 percent of the load existing prior to the creation of the DCAP.

Approximately two-thirds of the phosphorus load to the Cannonsville Reservoir is estimated to originate from farming. Animal manure is by far the largest single source of phosphorus in the reservoir basin. On-site wastewater (septic systems) and urban areas are estimated to contribute only about 6 percent of the total. Accordingly, Delaware County has specified operational targets of reductions from farms, and septic systems and urban areas combined, as 8,000 kilograms and 2,000 kilograms respectively. Options for achieving significant reductions in phosphorus from septic systems and the urban areas are limited. Therefore reducing the phosphorus conveyed by animal manure is critical for Delaware County, and for New York City, to reduce the overall loading of phosphorus sufficiently to securely restore the Cannonsville Reservoir (Cerosaletti *et al.*, 2004). Specific components of the DCAP designed to meet the target goals are described below.

*Animal manure and farm nutrient management*

Scientists from Cornell University and Cornell Cooperative Extension created nutrient management methods and tested these on pilot farms in the Cannonsville Reservoir basin. Results showed as much as two-thirds of the nutrients imported annually onto farms remained as surplus. These annual surpluses greatly increased risks of elevated concentrations of phosphorus in run-off (Sharpley, 2000). The fieldwork showed that precision feeding for animals could greatly reduce this risk by reducing the amount of phosphorus needed in production (Cerosaletti *et al.*, 2004).

Precision feed management is a truly comprehensive form of nutrient management central to the farm mass nutrient balance. Purchased feeds normally account for 65 to 85 percent of the nutrients imported onto a farm. By precision feeding the inputs can be reduced by more than 30 percent without losses in animal production. Precision feeding is achieved by careful regulation of the amounts of nutrients fed to the animals to precisely and quantifiably match their nutrient requirements. It is estimated that if all dairy animals in the Cannonsville watershed were fed according to precision feeding methods, the reduction in phosphorus on the farms in total could be up to 75,000 kilograms per year.

Losses of phosphorus from farms can be further reduced by improving the quality of on-farm forage. This consumes a larger fraction of the soil phosphorus thereby reducing losses in run-off. These reductions in annual phosphorus levels on farms are achieved while sustaining or increasing farm production. Farmers therefore have an economic incentive to adopt the methods. Complementary research focused on reliable and practical methods for the farmer to improve the use of manure on fields and crops while protecting water quality (Kleinman *et al.*, 1999). An overview of the scientific work identifying and assessing farm management options of phosphorus is presented in Porter *et al.* (2008).

## Community planning, stormwater and highway drainage

The Delaware County Department of Planning now explicitly takes into account water quality aims in all of its procedures and activities. Its planning procedures include comprehensive community planning, environmental impact assessments, zoning, subdivision regulations, site plan reviews, and planning to protect local water supplies. The planning department assists communities in the county through its Town Planning Advisory Service (TPAS). Of particular importance is the Delaware County Highway Management Plan that the planning department instituted in conjunction with the County Department of Public Works. This consists of the following:

1   Inventory and evaluation of road conditions and formulation of appropriate maintenance programmes for county and town roads referred to as Highway Management Plans.
2   Evaluation of all roads and their drainage as conduits of pollutant delivery.
3   Implementing stormwater management practices (SMPs) in areas identified in the evaluation phase as significant pollutant contributors.
4   Fostering an institutional relationship and cooperative partnerships with the towns to ensure the continued implementation of the maintenance programmes and SMPs.

Highways capture drainage from surrounding land as well as from the impervious surfaces of the highways. Thus highways are the major conduit of watershed drainage (Hoang and Porter, 2010). The greatest impact of non-point pollutant sources from fields and villages on water quality occurs through stormwater, as impervious areas and roads and their ditches are significant sources of contaminants to watercourses. Maintenance and repairs include sediment removal from the culverts, maintenance of catch basins and ditches, culvert repairs, and improved highway de-icing methods in winter.

*Solid waste management center and compost facility*

The Department of Public Works manages a state-of-the-art county-funded composting facility. Completed at an estimated cost of about US $20 million, the facility allows most of the total waste stream in Delaware County to be composted and recycled. The facility treats wastewater treatment biosolids, including animal manure, and commercial waste organic materials, and comprises a complete containment system for solid and hazardous wastes. This treatment produces compost as a commercially valuable product. Combined with an annual "clean sweep" for hazardous materials, the overall benefit of the facility is a very substantial reduction in risks to ground and surface waters in the county. The annual comprehensive "clean sweep" programme to collect wastes including pesticides, corrosives, solvents, and other harmful materials is itself a most effective way to reduce environmental risks.

*On-site wastewater treatment*

The Delaware County Soil and Water Conservation District assessed on-site wastewater treatment systems throughout the Cannonsville basin according to soil and hydrological conditions (Day, 2001). There are about 7,000 on-site septic systems in the basin. Rehabilitation or replacement of the majority of these systems was found to be desirable because of their average age (30 years), their outdated designs, and the unfavorable drainage of soils in which they were placed. A comprehensive rehabilitation and maintenance programme is now managed through the Catskill Watershed Corporation with funding provided by New York City (Catskill Watershed Corporation, 2013).

*Stream corridor protection and rehabilitation*

The west branch of the Delaware River includes almost 1,070 kilometers of streams above the Cannonsville Reservoir. Stream corridors are the last barrier to non-point source contaminants draining from the land. Working with local residents, landowners, and municipalities the Delaware County Soil and Water District has assessed and mapped the stream corridors. Protection measures are planned and implemented through the Stream Corridor Management Plan according to priorities and available funds. The programme is implemented with the New York City Department of Environmental Protection (NYCDEP) and US Army Corps of Engineers. These agencies have provided well over US $3 million for the work. The USEPA requires this programme under its filtration avoidance agreement with New York City. Providing guidance to local agencies and landowners, the programme uses the "natural stream channel" design concept for stream bank and channel stabilization, water quality improvement, aquatic habitat enhancement, and flood mitigation potential. Apart from flood control the benefits include:

- protection of agricultural land and public infrastructure;
- increased stream reach stability;
- improved aquatic habitat;
- reduced turbidity, sedimentation, and associated nutrient loading;
- filtration avoidance;
- community and interagency participation in solving a complex issue;
- expedited permitting for municipalities; and
- improved habitat for upland wildlife.

*Evaluation and monitoring supported by modelling and research*

As noted, a core principle is that the DCAP is founded on sound scientific credentials. The DCAP created its Scientific Support Group (SSG) with the goals of: providing a scientific basis for assessing management needs and options in the watershed; verifying scientific validity of management measures; and subsequently evaluating management options that are implemented. Quantifying non-point sources and the benefits of management measures can be difficult, posing significant challenges for monitoring in particular (Novotny, 2003). Until recently, there were six full-time water quality monitoring stations in the Cannonsville Reservoir basin (Bishop, 2004). The data obtained have supported exhaustive mathematical and statistical analyses combined with other scientific field studies. Mathematical modelling by Cornell University scientists supports management decision-making and evaluation of decisions implemented in the watershed (Tolson and Shoemaker, 2004).

*Economic development*

In accord with the MOA, the DCAP assumes that protection of water quality and economic development are compatible. It seeks to integrate source water protection with the economic and social goals of the communities concerned. Thus the Delaware County Department of Economic Development promotes:

- traditional economic and industrial development in Delaware County outside the New York City Watershed;
- environmentally benign initiatives with communities and businesses within the watershed.

The Department provides funds, information services, and counselling to businesses. An example is the Main Street Revitalization Program that integrates environmental, amenity and economic benefits to streets in the communities.

## Getting things done

The core agencies of the DCAP are the Delaware County Departments: Economic Development, Planning, Public Works, Soil and Water Conservation

District, and the Office of Watershed Affairs. It is a successful locally driven programme directly accountable to local elected officials. Inclusion of the Economic Development and Planning departments recognizes that local planning is a key tool in ensuring adequate source water protection. The staff members of the several county departments serve as highly professional technical providers and are well-qualified, trained, and experienced. They are locally known, respected, and accepted by landowners, residents, and other stakeholders in the watershed, which is essential for their work and roles as advisers, intermediaries, and enablers to be successful.

Although the work of the DCAP is locally led there are also close working collaborative arrangements with all the key watershed partners (see Table 4.1). These include the Catskill Watershed Corporation and city, state, and federal agencies. A hierarchy of interagency committees supported the work in its launching. For example, the Scientific Support Group has guided the scientific work; a Phosphorus Study Committee provided an interagency review of progress; and a Water Quality Committee comprising town supervisors reviews and decides policies. The county departments themselves meet frequently to coordinate and review their respective activities in the DCAP Core Group meetings. Taken together, this complex of cooperative partnerships provided a leading example of environmental management achieved through multi-level governance and interagency collaboration.

*Table 4.1* DCAP partners

| *DCAP partners* | |
| --- | --- |
| Local | • Delaware County: Planning Board, Office of Economic Development, Department of Public Works, Chamber of Commerce, Industrial Development Agency, Soil and Water Conservation District. |
| | • Other: Cornell Cooperative Extension, New York/ Delaware County Farm Bureau, Coalition of Watershed Towns, community organizations. |
| Regional (New York City Watershed) | • Catskill Watershed Corporation, New York City Department of Environmental Protection, Watershed Agricultural Council. |
| State (New York) | • New York State Departments of: Environmental Conservation, Health, State, Transport, Agriculture and Markets. |
| | • Other: NYS Water Resources Institute, Cornell University, Syracuse University, New York State Soil and Water Conservation Committee, NYS Conservation District Employees Association. |
| Federal | • US Army Corps of Engineers. |
| | • Environmental Protection Agency. |
| | • United States Department of Agriculture: Natural Resources Conservation Service, Agricultural Research Service, Farm Services Agency. |

Source: Delaware County Department of Watershed Affairs

Water source protection initiatives in the New York City Watershed, and particularly agricultural programmes and measures, have benefited from a level of funding from the City that may not be readily obtainable elsewhere because of the fiscal incentive provided by filtration avoidance. However, the acceptance of responsibilities for watershed management by local government means that locally generated finance for measures has also been very significant, and actions and local investments under the DCAP exemplify this. Finally, the well-developed arrangements for cooperation and collaboration between organizations and levels of government mean that both existing agency budgets and available state and federal grants have been successfully bid for and employed.

## Outcomes

When the DCAP was created in 1998, the Cannonsville Reservoir basin was "phosphorus restricted." This restriction was due to the level of phosphorus in the reservoir then exceeding the guidance value of 20 μg/L. That level has now dropped to about 17 μg/ L and the "restricted" designation has been lifted. As a critical consequence of this, new discharges of phosphorus can be permitted providing for economic growth. This outcome is at least partially attributable to the DCAP's comprehensive planning and management of all significant non-point sources within the Delaware County region of the New York City Watershed. Planning and identification of management options are determined and evaluated with sound scientific credentials. The scientific work conducted in the Cannonsville Reservoir watershed, with overall total funding now exceeding US $20 million, has established an understanding of the cause and effects of non-point sources that is probably unique globally in its range and depth. Since the DCAP was established over US $10 million has been raised through external funding for studies, planning, and implementation. These funds have supported advanced nutrient management on farms, highway drainage, and stormwater and flood protection.

The DCAP and the combined efforts of the Catskill Watershed Corporation (CWC) and the NYCDEP demonstrate that sanitary management of multiple non-point sources in an inhabited watershed can provide a viable and sustainable source of high-quality drinking water. This management balances regulations imposed through higher police powers administered through the NYCDEP, and voluntary acceptance of management responsibilities at the local level fostered by the DCAP and the CWC. These watershed partners seek to promote the view that the Watershed Rules and Regulations should be a "helping" rather than "heavy" hand. The New York City Watershed represents rigorously comprehensive protection of a water supply through source water protection. Such protection is an alternative to very expensive filters otherwise required under federal law. A high proportion of the watershed is open space and forest and this yields high-quality water (Dombeck, 2003), but it is also a "living watershed" with farms, commercial, residential, and other land uses. Integrating environmental and economic objectives and sustaining water quality has been achieved through the necessary local acceptance of responsibility and the development of local capabilities, with support and guidance from scientific partners and higher levels of government.

## Concluding reflections

### *Institutional, technical and political challenges*

Although much institutional and operational strength can be identified there are also unresolved institutional and technical challenges in the New York City Watershed. Despite the acknowledged success of the DCAP it is not generally included as an equal partner by state agencies, New York City, and the USEPA when major questions about the watershed are being decided. The board of supervisors in Delaware County regards this at least partial exclusion of local partners from being a party to such negotiations as contrary to the spirit and letter of the MOA.

It is relevant to note that there is no overall governance of the Catskill–Delaware watershed region that is answerable or accountable to local constituencies. New York City, through its Department of Environmental Protection, governs the watershed under the authority of its Watershed Rules and Regulations as approved by the New York State Department of Health. However, residents in the watershed have no electoral voice in the government of New York City. To that extent the watershed residents and communities are disenfranchised. The Catskill Watershed Corporation is accountable to the local electorate in the watershed through its board members who are locally elected officials. However, the CWC has no governance authority and serves primarily as a funding body. The absence of a regional watershed entity was a matter of choice on the part of the watershed municipalities who had no wish for a regional authority to be established as part of the MOA; reflecting the disfavor with which regional government tends to be viewed in the United States (Negro and Porter, 2009).[2] This means that watershed governance that is accountable to the local electorate must be applied through the existing municipal structure such as counties. This is the institutional role the Delaware County Action Plan has sought to accomplish.

One key issue has been particularly contentious. This is the question of how much land New York City should acquire in the watershed to maintain as unoccupied open space to ensure the sustainable security of its source waters. The question equates to: what is sufficient protection and how to achieve it? Leaders in the environmental community involved with watershed affairs favor one view in particular.

> When land use controls become necessary to protect the environment, frequently nothing less than a complete ban on further development can ensure preservation. The legal issues then posed do not involve how much development is permissible, but whether development is permissible at all.
>
> (Malone, 2001, p. xi)

Economic development is the underlying major point of contention in the New York City Watershed. Although the MOA clearly affirms the dual goals of protection for the water supply and maintaining the economic vitality of watershed communities, some prefer the latter to be more aspiration than achieved in practice. Some leading members of the environmental community believe promotion of economic interests of the watershed communities to be unsafe for water quality. For example, in a critique of the MOA, Goldstein (2001) argued that the government agencies needed to recognize that "Mother Nature knows best." Goldstein emphatically argued that watershed protection is best accomplished by curtailing human activities in the watershed. Carried to its logical conclusion, ultimate protection means preserving a wilderness. The influential conviction that protection means preserving "the wilderness" is thoroughly discussed for the regional context by Oelshlaeger (1991), and the efforts of New York City were described in an article in *Nature* as restoring "the integrity of the Catskill ecosystems" (Chichilnisky and Heal, 1998). This position fails to recognize that faithful restoration of the integrity of the Catskill's ecosystem would require eliminating the huge reservoirs by the demolition of the dams that have permanently submerged about 22,600 acres or 35.3 square miles of the watershed. To local leaders in the watershed communities this view equates to New York City being free to damage the environment for its own development goals while denying the watershed communities their own interests in development.

In 2007, the EPA granted to New York City a new filtration avoidance determination (FAD) for the water supply system. A particularly contentious requirement in this new FAD was that New York City should allocate US $300 million for a ten-year Land Acquisition Program (LAP) in the watershed. Land acquisition had been a major part of previous FADs. However, the scale and scope of the new LAP was far beyond that previously envisaged. Delaware County, as the county with more than half of the land area comprising the Catskill–Delaware watershed, was alarmed by the economic and social consequences that a US $300 million land acquisition programme represented.

There were two primary objections to the LAP as proposed. First, the planned land purchase and conservation easement programme could permanently remove from development much of the land that is currently developable in the county's watershed area. Delaware County, as previously mentioned, consists largely of steep-sided hills and narrow valleys. The land potentially available for new development is therefore limited. A land purchase programme on the scale required by the USEPA and New York State in the current FAD could render much of the county non-viable in terms of future economic growth. A second major concern was that local governments rely upon property tax revenues as their primary revenue source. Land permanently taken out of development and maintained as open space represents a much lower tax base than developed land. Hence future tax revenues for communities in the county could be severely impaired. A third objection voiced by Delaware County was that no consideration appeared to have been given to alternative uses of US $300 million. Such a large sum could fund other effective means of reducing potential pollution.

Three years of intense and difficult negotiation followed the 2007 FAD as New York City sought a permit from the state to acquire land as required by the FAD. On February 16, 2011, the NYS Department of Environmental Conservation (NYSDEC) and the NYCDEP jointly announced agreement on a new fifteen-year Water Supply Permit for New York City (NYSDEC, 2010). This agreement authorizes New York City to acquire up to 105,000 acres of land in the Catskill–Delaware watershed. As a result, the percentage of watershed land owned by the state and the City will increase from just over 33 percent to 44 percent. Strong incentives to support this agreement were also provided to watershed communities. These incentives included: continued funding for water quality protection and on-site wastewater treatment; the provision of enhanced recreational opportunities on land owned by the City; authorization for watershed towns to exempt specified lands from acquisition by the City to allow for economic development in the towns; and assurance that property taxes owed by the city to watershed municipalities would be fairly calculated. With such concessions, the Coalition of Watershed Towns and its constituents agreed to support the new permit for land acquisition.

### Water treatment versus watershed protection: an ongoing question

Following the passage of the Safe Drinking Water Act Amendments in 1986, the chief of the Drinking Water Quality Control Division of the New York City Department of Environmental Protection suggested that New York City could take one of two directions in sustaining high-quality water supplies. It could completely depend on water treatment technology with its "associated economic and technical responsibilities of unimaginable magnitude" or it could take a second direction:

> The alternative direction, characterized by a policy of resource protection and scientific surveillance, affords advantages in maintaining a cost-effective, high-quality drinking water without the complexities of superfluous treatment technology and source quality degradation. New York, because of its unique high-quality surface supply, again has the opportunity to lead in establishing principles for drinking water preservation that are consistent with the lessons of its historical supply development. The principles of surface water preference, a remote supply, sanitary protection, maximum utilization of natural quality, disinfection, scientific surveillance, and selective application of technology can be synthesized into a policy of reservoir protection that has utility for filtered and non-filtered sources alike.
>
> (Iwan, 1987, p. 203)

The New York City Watershed communities are committed to this alternative direction. The watershed is a "protected area," but it is also a "living landscape" whose protection depends upon:

> a vigorous economy and social structure, and a population that is sympathetic to the objectives of conservation. This necessitates working with people at all levels and especially with those who live and work in the area—the people most intimately affected by what happens to it.
>
> (Lucas, 1992, p. xvi)

The New York City Watershed safeguards the source waters through protection achieved by high sanitary standards. Such safeguarding requires communities, landowners, and residents in the watershed to willingly accept local responsibilities. This is a challenge Delaware County, with its watershed partners, is meeting through its locally driven Delaware County Action Plan.

In this protection, the DCAP and the CWC demonstrate effective "home rule" for watershed management. Such watershed organization is a new paradigm based on systematically managing non-point sources to achieve that protection through measurable water quality objectives. At the same time it seeks to promote the economic viability of the communities it serves through environmentally benign economic development. Highly professional local technical providers represented by the core agencies of the DCAP and the CWC's staff serve these dual aims. John Nolon (2003) has noted the trend shown by local governments throughout the United States in adopting an impressive number of local environmental laws. Delaware County, through the DCAP, and the entire Catskill–Delaware watershed through the CWC, are bridging the gaps between federal and state law, and local law related to land and water uses (Porter, 2005). The programme works through active partnerships between different levels of authorities together engaged in the New York City Watershed, and represents a remarkable voluntary assertion of environmental responsibility by local government leaders in a county and region that has extremely limited economic resources. This collaborative programme is a nationally significant demonstration of protection at its source of a water supply for nine million people. It is an innovative demonstration of willingness to meet the costs of watershed protection by those who benefit from it, offering value for money to the water consumers of New York City.

## Acknowledgments

Particular thanks are owed to Dean Frazier and his colleagues in Delaware County for sharing their time and expertise with the authors.

## Notes

1   Detailed maps, description and history of the New York City (NYC) water supply system can be viewed at the website of the NYC Department of Environmental Protection (http://www.nyc.gov/html/dep/html/drinking_water/wsmaps_wide. shtml).

2   The 1997 MOA did additionally create a Watershed Protection and Partnership Council (WPPC) to function as a forum for the exchange of ideas and information regarding watershed protection and economic growth in the upstate communities. An aim was to help with dispute resolution and cooperation. The WPPC does not have regulatory power or enforcement capabilities, but is charged with periodically reviewing governmental and private efforts to protect the watershed and soliciting of input from governmental agencies, private organizations, and other stakeholders. Unfortunately, the WPPC has not successfully involved all interested stakeholders in managing the watershed. While the New York City Watershed is regional in scale, the WPPC lacks the authority, legitimacy, and means to independently implement a regional water management strategy and integrate the City with the upstate stakeholders.

## References

Ashendorff, A., Principe, M. A., Seeley, A., LaDuca, J., Beckhardt, L., Faber Jr., W. and Mantus, J. (1997) 'Watershed protection for New York City's supply', *Journal of American Water Works Association*, vol. 89, no. 3, pp. 75–88.

Bishop, P. (2004) *Water Quality Monitoring in the Cannonsville Reservoir Watershed*, RELU Meeting, Cornell University, October 25, 2004, www.inmpwt.cce.cornell.edu/documents/CannonsvillePat%20Bishop10-25-04.pdf, accessed September 16, 2013.

Catskill Watershed Corporation (2013) *Septic Programs*, www.cwconline.org/cwc_website_new_023.htm, accessed September 16, 2013.

Cerosaletti, P. E., Fox, D. G. and Chase, L. E. (2004) 'Phosphorus reduction through precision feeding of dairy cattle', *Journal of Dairy Science*, vol. 87, no. 7, pp. 2314–2323.

Chichilnisky, G. and Heal, G. (1998) 'Economic returns from the biosphere', *Nature*, vol. 391, pp. 629–630.

Committee on Public Health (1989) 'Statement on Preservation of New York City's Drinking Water Quality', *Bulletin of New York Academy of Medicine*, vol. 65, no. 8, pp. 898–904.

Committee to Review the New York City Watershed Management Strategy (2000) *Watershed Management for Potable Water Supply: Assessing the New York City Strategy*, National Academy Press, Washington, DC.

Day, L. D. (2001) *Phosphorus Impacts from Onsite Septic Systems to Surface Waters in the Cannonsville Reservoir Basin*, NY, Delaware County Soil and Water Conservation District, Walton, New York.

Delaware County Department of Watershed Affairs (2002) *Delaware County Action Plan (DCAP II) for Watershed Protection and Economic Vitality*, Delaware County Department of Watershed Affairs, Delhi, NY.

Dombeck, M. (2003) 'From the Forest to the Faucet', in McDonald, B. and Jehl, D. (eds) *Whose Water Is It?* National Geographic Society, Washington, DC.

Galusha, D. (1999) *Liquid Assets: A History of New York's Water System*, Purple Mountain Press Ltd., Fleischmanns, NY.

Goldstein, E. (2001) 'Mother Nature knows best: Fundamentals for ensuring a safe water supply', *Fordham Environmental Law Journal*, vol. 12, pp. 455–465.

Hoang, Y. and Porter, K. S. (2010) 'Stormwater management in the rural New York headwater areas of the Chesapeake Bay watershed', *The Journal of Water Law*, vol. 21, no. 1, pp. 6–18.

Hu, W. (2004) 'Water, and competing rural interests, everywhere: Striving to protect the watershed, the city assumes the role of country land baron', *New York Times*, August 9.

Iwan, G. R. (1987) 'Drinking water quality concerns of New York City, past and present', *Annals of the New York Academy of Sciences*, vol. 502, pp. 183–204.

Kavanaugh, J. (1999) 'To filter or not to filter: A discussion and analysis of the Massachusetts filtration conflict in the context of the Safe Drinking Water Act', *Boston College Environmental Affairs Law Review*, vol. 26, no. 3, pp. 809–829.

Kelly, W. J. (1986) *The Safe Drinking Water Act Amendments of 1986*, BNA Special Report, Bureau of National Affairs, Washington, DC.

Kennedy Jr., R. F. (1997) 'A Culture of mismanagement: Environmental protection and enforcement at the New York City Department of Environmental Protection', *Pace Environmental Law Review*, vol. 15, no. 1, pp. 233–254.

Kleinman, P. J. A., Bryant, R. B. and Reid, W. S. (1999) 'Development of pedotransfer functions to quantify phosphorus saturation of agricultural soils', *Journal of Environmental Quality*, vol. 28, no. 6, pp. 2026–2030.

Longabucco, P. and Rafferty, M. (1998) 'Analysis of material loading to Cannonsville Reservoir: Advantages of event-based sampling', *Lake and Reservoir Management*, vol. 14, no. 2–3, pp. 197–212.

Lucas, P.H.C. (1992) *Protected Landscapes: A Guide for Policy-makers and Planners*, Springer, London.

Malone, L. (2001) *Environmental Regulation of Land Use*, West Group, St. Paul, MN.

Negro, S. and Porter, K. S. (2009) 'Water stress in New York State: The regional imperative?', *The Journal of Water Law*, vol. 20, pp. 5–16.

Nolon, J. R. (2003) 'In praise of parochialism: The advent of local environmental law', in Nolon, J. R. (ed.) *New Ground: The Advent of Local Environmental Law*, Environmental Law Institute, Washington, DC.

Novotny, V. (2003) *Water Quality: Diffuse Pollution and Watershed Management*, J. Wiley, Hoboken, NJ.

NYSDEC (2010) *NYC Watershed Land Acquisition Program*, www.dec.ny.gov/permits/70376. html, accessed September 16, 2013.

Oelshlaeger, M. (1991) *The Idea of Wilderness*, Yale University Press, New Haven, Connecticut.

Okun, D. A., Craun, G. F., Edzwald, J. K., Gilbert, J., Pannetier, E. and Rose, J. B. (1993) *Report of the Expert Panel on New York City's Water Supply*, USEPA, Washington, DC.

Okun, D. A., Craun, G. F., Edzwald, J. K., Gilbert, J. and Rose, J. B. (1997) 'New York City: To filter or not to filter?', *Journal of American Water Works Association*, vol. 89, no. 3, pp. 62–74.

Porter, K. S. (2005) 'Should governmental water responsibilities flow downwards?', *The Journal of Water Law*, vol. 16, no. 2, pp. 49–57.

Porter, K. S., Porter, M. J., Bishop, P., Pacenka, S. and Tsoi, S. I. (2008) *A Compendium of Scientific Work on a New York State Agricultural Watershed*, New York State Water Resources Institute, www.delcowatershed.com/, accessed September 4, 2013.

Rodenhausen, G. A. (2001) 'Water supply and stream protection' in Ginsberg, W. R. and Weinberg, P. (eds) *Environmental Law and Regulation in New York*, Thomson West, Eagan, MN.

Sharpley, A. N. (2000) 'Concluding remarks: Future strategies to meet the agricultural and environmental challenges of the 21st century' in Sharpley, A. N. (ed.) *Agriculture and Phosphorus Management*, Lewis Publishers, Boca Raton, FL.

Tolson, B. and Shoemaker, C. (2004) *Watershed Modeling of the Cannonsville Basin using SWAT 2000*, Technical Report, School of Civil and Environmental Engineering, Cornell University, www.swat.tamu.edu/media/1328/cannon_report1_0s.pdf, accessed September 16, 2013.

# 5 The Hudson River Watershed, New York State, USA

*Mary Jane Porter, Keith Porter and Laurence Smith*

## Introduction

The Hudson River is a defining natural feature of a major region of New York State. From New York City north to Albany, the Hudson River Valley is a National Heritage Area. "The Hudson is one of North America's most important, best-studied, and fiercely protected rivers" (Benke and Cushing, 2005, p. 35). Its greater watershed encompasses almost 35,000 square kilometers, including parts of four other states: Vermont, Massachusetts, New Jersey, and Connecticut. The many tributaries in the entire Hudson River Watershed themselves define watersheds. Rising in the Adirondacks the river flows for over 500 kilometers to the tip of Manhattan island.[1] Almost half of this length, the 246 kilometers from Troy Dam south to New York Harbor, is the Hudson River Estuary. This is tidal and drains approximately 13,500 square kilometers. With the rising and falling of the tide in the estuary come changes in the direction of flow. Generally speaking, a rising tide is accompanied by a *flood* current flowing north towards Troy and a falling tide by an *ebb* current flowing south towards the sea. Sea water pushing up the estuary is diluted by freshwater runoff flowing south. Thus the northerly limit of the "salt front" of diluted sea water varies seasonally with the amount of precipitation in the watershed. Upstream of the dam at Troy the river is no longer influenced by the tide and thus not categorized as an estuary.

Estuaries are highly productive and diverse ecosystems and more than 200 species of fish are found in the Hudson and its tributaries (Benke and Cushing, 2005). This productivity is ecologically and economically valuable to much of the Atlantic coast as well as to the estuary itself. Both commercial and recreational fisheries depend on the nursery habitats of the estuary, while tidal marshes, mudflats, and other significant habitats in and along its length support a great diversity of birds and other wildlife, including endangered species such as the shortnose sturgeon.

Although the Hudson's watershed may be considered densely populated as nearly one-half of New York State's population lives in the 15 counties bordering the estuary, it contains a mosaic of densely settled urban areas, less densely populated suburbs and small towns, and large expanses of forested and other open landscapes; including the Catskill and Adirondack forests. Uses of the estuary include commercial navigation, recreation (fishing, swimming, boating, and wildlife observation), and commercial fishing, as well as municipal drinking water supplies. Located along the river are major power-generating facilities, manufacturing plants, petroleum terminals, cement and aggregate plants, resource recovery facilities, and various mining operations (NYSDEC, 2010).

In the past the Hudson's natural resources were abused as population, agriculture, and industrialization all increased. Discharges of raw sewage led to high bacterial counts and low oxygen levels, wetlands were destroyed by landfilling, scenic vistas spoilt by industry, fish killed by cooling water intakes, and food webs contaminated by toxic chemicals. But in recent years many citizen groups and government agencies have worked actively to restore and protect the estuary's natural resources. Today the Hudson's waters flow cleaner than they have in decades. It has become one of the healthiest estuaries on the Atlantic coast of the USA and once more attractive for recreation and wildlife. Revitalized public beaches, boat launches, parks, and angling access points now abound on the estuary's shores (NYSDEC, 2012).

The Hudson Valley critically needs integrated and effective watershed governance to deliver such improvements, and to face future challenges. Throughout the region there is increasing demand for water, an ageing and often failing infrastructure and potential impacts of climate change. During the next inevitable drought, daily demand for water in the region, including New York City, could greatly exceed the available supplies. For example, during the 1985 drought, demand exceeded the system's safe supply by over 750,000 cubic meters per day (Padavan, 1988). Although New York City has subsequently greatly improved the efficiency of its water supplies it is foreseeable that the region would be very seriously challenged during a severe drought to meet a daily demand that has greatly increased since the 1980s.

It is also anticipated that climate change will amplify stresses on the region's water resources. In 2001 Rosenzweig and Solecki warned that through precipitation, temperature, and its effects on evapotranspiration, sea-level rise climate change would almost certainly have effects on Metro East Coast water systems (Rosenzweig and Solecki, 2001). In addition, development in the watersheds stresses the region's water supply reservoirs by adding pollutants or reducing the filtration provided by forests and wetlands (Gordon and Kennedy, 1991). Urban development also covers increased areas of land with impervious surfaces that water cannot penetrate. The resulting greater volumes of stormwater, rather than recharging groundwater, also become an increased source of pollutants. In addition to adverse impacts on the quality of the valley's water resources, stormwater runoff has also caused more frequent flash flooding throughout the region in recent years (Foderaro, 2007).

This chapter describes the collaborative and integrated governance arrangements and improvement programmes that have developed in the Hudson Valley. Its specific focus in keeping with the themes of this book is the Hudson River Estuary Program (HREP) and its fostering of management of tributary watersheds through intermunicipal agreements in the Hudson Valley. The HREP was created in 1987 and aims to protect and improve the natural and scenic Hudson River Watershed for all its residents. Its core mission is to:

- ensure clean water;
- protect and restore fish, wildlife and their habitats;
- provide water recreation and river access;
- adapt to climate change;
- conserve the world-famous scenery.

Guided by a forward-looking plan developed through extensive community participation the HREP achieves its aims through outreach, coordination with state and federal agencies and public–private partnerships. This collaborative approach includes grants and restoration projects, education, research and training, natural resource conservation and protection, and community planning assistance. Projects are founded in science and carried out in ways that support the citizens and communities (NYSDEC, 2012).

## Context and drivers

### Environmental activism

Dismayed at the abuses of the Hudson River, citizens started to take action. Early action in the late nineteenth century saw New York and New Jersey residents combining to mount a preservation effort that saved the Palisades cliffs. Then over the last five decades several disputes have arisen over uses of land adjacent to the river and uses of the river itself which have had the potential for water pollution and ecosystem degradation. In 1962 a "landmark" dispute between a small group of conservationists and a powerful utility took place over a proposed power plant on the river's Storm King Mountain. The proposal for this plant began a seventeen-year legal battle that helped to found national environmental activism, setting far-reaching precedents for conservation in America and globally, including establishment of the right of citizen groups to sue a government agency to protect natural resources (Dunwell, 2008).

Three conservationist groups voicing their concerns and planning action about the continued degradation of the river and the proposed power plant at that time were Scenic Hudson, Riverkeeper, and the Hudson River Sloop *Clearwater*. Campaigning in the Hudson Valley since 1963, Scenic Hudson has been credited with saving Storm King Mountain and "launching the modern grass-roots environmental movement" (Scenic Hudson, 2013). Their mission is "to protect and restore the Hudson River and its majestic landscape as an irreplaceable national treasure and a vital resource for residents and visitors" (Scenic Hudson, 2013). Guiding principles are that: outstanding quality of life is achievable only when a clean, healthy environment is a key component of economic development; all citizens have a right to outstanding quality of life, including access to our Hudson River, to open space and to participate in community decision-making; our natural environment is an irreplaceable source of spiritual and artistic vitality and must be preserved forever (Scenic Hudson, 2013).

Today Scenic Hudson has over 25,000 supporters and promotes land preservation by creating or enhancing natural parks, and through partnerships with farmers and other private landowners in its stewardship programmes. For example, a recent grant from a charitable foundation will be used to define specific priorities for protecting farmland on a "parcel-by-parcel" scale and to research and promote the capacity of municipal and county governments to invest in keeping land in agriculture, fostering farmland preservation collaborations between municipalities and land trusts, state and federal government, and private donors. To date US $19.5 million has been invested in working with farm families to conserve and enhance their operations to provide healthy, local food and sustain small businesses in the regional economy. Scenic Hudson also works with communities in revitalizing riverfronts and through volunteer programmes and environmental education for citizens (Scenic Hudson, 2013).

A group of commercial and recreational fishermen were also dismayed at the dying state of the Hudson in the mid-1960s, and in 1966 became the Hudson River Fishermen's Association (HRFA). They identified and brought action against polluting groups using a little known law—the Rivers and Harbors Appropriation Act of 1899—which in its Section 13 forbade disposal of refuse matter in any navigable waters in the United States. Financed, at least in part, by the bounties and awards from their lawsuits, the HRFA launched their first Riverkeeper patrol boat in 1983. Today Riverkeeper is a member-supported watchdog organization with the mission "to protect the environmental, recreational and commercial integrity of the Hudson River and its tributaries, and to safeguard the drinking water of nine million New York City and Hudson Valley residents" (Riverkeeper, 2013). It continues to patrol the Hudson River from Troy Dam to New York Harbor, challenging illegal and polluting activities, while also working to strengthen the regulations and laws that protect New York's water and waterways, and reaching out to the public to enlist their assistance through educational activities.

Similarly dismayed by the pollution of the Hudson, environmental activist and folk musician Pete Seeger announced plans in 1966 "to build a boat to save the river" (HRSC, 2013). The purpose was to assist people in experiencing its beauty and instill a desire to help in the river's protection, and in 1969, the sloop *Clearwater* was launched, a replica of a type commonly used in the eighteenth and nineteenth centuries on the Hudson River. It is noted for being one of the first US vessels "to conduct science-based environmental education aboard a sailing ship, virtually creating the template by which such programmes are conducted around the world today" (HRSC, 2013). The mission of the *Clearwater* "is to preserve and protect the Hudson River for the benefit of its ecosystem and human communities while creating new environmental leaders for a sustainable future." Their primary concerns include: investigating and conducting research in causes or sources of contamination and informing the public of such dangers; educating citizens concerning the importance of preserving the river; fostering its cultural and historic heritage; and conducting activities that enhance and improve the Hudson River's environment (HRSC, 2013). To date the *Clearwater* has introduced more than half a million young people and hundreds of thousands of adults to the Hudson River Estuary ecosystem. It has become an icon for the Hudson River, and was named on the National Register of Historic Places in 2004 for its environmental role.

Although Scenic Hudson, Riverkeeper, and the sloop *Clearwater* are leading and long-standing examples, they are by no means alone in their activism to restore and protect the Hudson and its watershed:

> The Hudson is home to one of the most vigilant, sophisticated, and aggressive environmental communities to protect any resource in the world. Today there is a citizen group at every bend and on both banks ready to do battle with any potential polluter or developer.
>
> Robert F. Kennedy, Jr., foreword to *The Hudson: America's River*
> (Dunwell, 2008, p. iv)

Indeed, a directory compiled in 2006 lists 51 environment-related organizations in the Hudson River Valley including government, non-profit, preservation/ conservation, and educational groups (Moira Productions, 2006).

### Institutional responses

The New York State Department of Environmental Conservation (NYSDEC) was created in 1970 to combine in a single agency all state programmes designed to protect and enhance the environment. Its mission is "to conserve, improve and protect New York's natural resources and environment and to prevent, abate and control water, land and air pollution, in order to enhance the health, safety and welfare of the people of the state and their overall economic and social well-being" (NYSDEC, 2014). It seeks to achieve this mission through the simultaneous pursuit of environmental quality, public health, economic prosperity, and social well-being, including environmental justice and the empowerment of individuals to participate in environmental decisions that affect their lives.

During the 1960s and 1970s public concern existed for the protection of the Hudson's fisheries. This led to the passage of the 1979 Hudson River Fisheries Management Act. Since then concerns have widened. Within the watershed aggressive land development has raised water stresses in times of drought and depletion of groundwater resources, while it contributes to damaging floods during storm events. These problems are exacerbated by the absence of regional instruments of governance. It was recognized that to conserve the river's fish, habitats, and ecosystems would require a broader and more multi-disciplinary approach, and in 1987, the fisheries law was replaced by the Hudson River Estuary Management Act (Section 11-0306 of the New York State Environmental Conservation Law). This act directed the NYSDEC to develop a management programme for the conservation of the estuary, including the tidal portion of the river as well as its shoreland. This programme became the Hudson River Estuary Program (HREP).

## Getting started

The Hudson River Estuary Management Act established the Hudson River estuarine district to include the tidal waters of the Hudson River, including the tidal waters of its tributaries and wetlands from the federal lock and dam at Troy to the Verrazano–Narrows Bridge at the mouth of New York Harbor. NYSDEC were required to establish a continuing fifteen-year Hudson River estuary management programme for this estuarine district and its associated shorelands in order to protect, preserve, and where possible, restore and enhance it. The commissioner of the NYSDEC was required to appoint a Hudson River estuary management committee to advise on regulation, policy, and the formulation of the programme consisting of at least 11 members representing estuary interest groups including commercial fishing, sport, research, conservation, and recreation. The commissioner was also required to maintain the post of Hudson River estuary coordinator to assist the commissioner and advisory committee in the development of the programme and to manage its implementation. In addition a Hudson River estuarine sanctuary was established to protect areas of special ecological significance and for research and education concerning the Hudson's ecosystem. This initially incorporated the existing four sites of the Hudson River national estuarine research reserve.

As defined in the Act the estuary management strategy was to include but not be limited to:

- characterization of how the estuary functions and the relative role of different species, with explanation of the estuary as a distinct ecosystem, a habitat for fish and wildlife, a commercial fishery and a recreational resource;
- identification of areas of special ecological significance and description of the annual dynamics of such habitats critical to habitat maintenance;
- a plan for the development and operation of the Hudson River estuarine sanctuary, including criteria for the identification of additional areas of special ecological significance and for education and research;
- a status report on the populations and relative abundance of species that have potential or existing recreational or commercial value or that play a key role in the functioning of the estuary and on the diversity of its species, and plans for annual monitoring;
- evaluation of the impact of water uses on the estuary;
- identification of areas of potential ecological significance requiring rehabilitation;

- a status report on toxicants and their effects on ecological indicator species, and species of commercial and recreational value;
- identification of the anthropogenic activities and the conservation and management problems posing a threat to the resources and functioning of the estuary;
- an inventory of ownership and tenancy of underwater lands in the estuarine district;
- recommendations for developing the economic potential of the Hudson River fishery and maintaining its traditional commercial fishery;
- recommendations for a fifteen-year estuary management programme;
- evaluation of the existing resources and authority of the NYSDEC to implement the estuary management programme, including research, information and data needs, and legislative, administrative, and regulatory recommendations, and the potential role of private sources and institutions.

The HREP Action Agenda 2010–2014 is the current guide for the estuary management programme initiated by this Act, and the result of a planning process that was launched in 1996 when the first Estuary Action Plan was approved. The Action Agenda was subsequently updated in 1998, 2001, and 2005, leading up to the 2010–2014 plan.

## Approach and tools

The HREP five-year Action Agenda was developed consultatively with communities along the length of the river. Implementation of the programme is achieved through extensive public outreach and voluntary action, coordination with state and federal agencies, and collaborative public–private partnerships with more than 500 partnering groups. Built on a sustained effort that identified the multiple interests of the estuary's citizens at the start of the programme, the resulting Action Agenda has widespread acceptance, and once adopted provides the basis for detailed work plans, budgets, and performance-based tracking systems. The collaborative approach includes:

- grants and restoration projects;
- education, research, and training;
- natural resource conservation and protection;
- community planning assistance.

The Action Agenda updates long-range goals and measurable targets for conserving, protecting, and revitalizing the Hudson River Estuary and its surrounding watershed. It is not a conventional state agency plan, but rather a statement and shared vision of where the state and its citizens in the Hudson Valley region want to be (NYSDEC, 2010). The goals and targets defined establish a framework for collaboration and recognize the critical roles that local governments, non-profit organizations, federal agencies, citizens groups, and a wide range of economic interests need to play to assure they are achieved. The reliance on partnerships is critical, recognizing that no single organization or agency can fund all of the science, education, and conservation initiatives needed to protect and restore the Hudson and its watershed.

The Action Agenda contains 12 long-range goals for the conservation and recovery of the Hudson Estuary and its watershed (Box 5.1). For each goal, actions to be completed by 2014 have been identified, as well as long-range targets up to 2020 or beyond. Some goals directly relate to one of five focus areas of the programme's mission, while others are cross-cutting and serve multiple aspects of that mission. Each programme staff member is responsible for particular goals in partnership with a "designated" goalkeeper who also assists in keeping the programme's goals on target. Progress on goals and their targets are reported on a timely basis through Report Cards and Progress Reports.

---

### Box 5.1  Hudson River Estuary Action Agenda goals, 2010–2014

1   Restore the signature fisheries of the estuary to their full potential, ensuring future generations the opportunity to make a seasonal living from the Hudson's bounty and to fish for recreation and consume their catch without concern for their health.

2   Conserve, protect, and enhance river and shoreline habitats to assure that life cycles of key species are supported for human enjoyment and to sustain a healthy ecosystem.

3   Conserve for future generations the rich diversity of plants, animals, and habitats that are key to the vitality, natural beauty, and environmental quality of the Hudson Valley.

4   Protect and restore the streams, their corridors, and the watersheds that replenish the estuary and nourish its web of life, and sustain water resources that are critical to the health and well-being of Hudson Valley residents and the ecosystem.

5   Conserve key elements of the working pastoral landscapes and world-famous river scenery that define the character of the Hudson River Valley, and provide new and enhanced vistas where residents and visitors can enjoy Hudson River views.

6   Address the causes of climate change in the Hudson Valley and prepare for projected impacts to safeguard our health and safety and to protect the natural resources and local economies that sustain our communities.

7   Develop, maintain and improve a regional system of access points for fishing, boating, swimming, hiking, education, river watching, and wildlife-related recreation, and build connections that allow residents and visitors to have rich and diverse river experiences.

---

8  Promote public understanding of the Hudson River, including the life it supports, its role in the global ecosystem, and the challenges the river faces and how they can be met.

9  Revitalize all the waterfronts of the valley so that the Hudson is once again the "front door" for river communities, where scenery and natural habitats combine with economic and cultural opportunity, public access, working ports and harbors, and lively adjacent downtowns, to sustain vital human population centers and a healthy environment.

10  Ensure that Hudson River water quality supports appropriate human benefits, including drinking water, swimming, fishing, navigation, and ecosystem protection.

11  Reduce contaminants entering the Hudson River, and remove or remediate river sediments contaminated by long-term pollutants so that food webs of the river are supported, people can safely eat Hudson River fish and harbors are free of the contaminants that constrain their operation.

12  Track our progress and celebrate our successes.

*Source: NYSDEC, 2010*

The HREP goals with most relevance to this book and comparability to its other case studies are numbers 4 and 10 in Box 5.1. Details of these goals and the action plans to achieve them are summarized below. The other ten goals provide examples of the diverse needs and actions required of a comprehensive and integrated watershed management programme on this scale. Of particular note are goals 8 and 12 in Box 5.1. These are cross-cutting goals and facilitating actions that are relevant for all watershed management programmes. Scientific assessments, decision support tools, monitoring and outreach, educational and cultural programmes are all key enablers for more direct measures such as stream bank restoration or provision of recreational access points.

### Action Agenda Goal 4

*Protection and restoration of streams, their corridor, and the watersheds that replenish the estuary and sustain Hudson Valley water resources*

This goal presents a challenge at the scale of the whole watershed and its landscape. The water quality and ecological health of the Hudson River is affected by what it receives from the upper Hudson and Mohawk rivers, as well as its lower tributaries. The Hudson is fed by thousands of miles of headwater streams and rivers that drain a populated and changing landscape while providing drinking water for millions of people, wildlife habitat, and recreation. Via these tributaries the watershed contributes fresh water and nutrients to sustain the estuary and its food chains. The confluences between the river and its tributaries also provide sites for important habitat and aquatic vegetation that support a variety of fish and wildlife. A healthy estuary thus requires a healthy watershed with intact riparian corridors, floodplains, wetland complexes, groundwater recharge, and minimal in-stream barriers. A high level of such watershed protection will also advance other environmental goals such as conserving habitat and providing resilience to climate change.

The Hudson's tributaries are impacted by a diversity of threats including increased water abstraction, increase in impervious surfaces created through land development, loss of ground-cover vegetation, agricultural and lawn runoff, failing wastewater treatment and septic systems, fish barriers, and atmospheric deposition of pollutants (NYSDEC, 2010). Given the desirable nature of the Hudson Valley as a location in which to live, economic development and its associated sprawl is a particular challenge for the environment. These threats degrade streams and rivers through erosion and siltation, polluted runoff, flooding, reduced groundwater recharge and low stream flows. Because the impacts can travel downstream and cross municipal boundaries, tributary health becomes a local and a regional issue.

The HREP aims to protect healthy streams before they become degraded and restore those stream corridors that have been damaged. To do this it promotes a locally led approach to watershed planning and conservation that is holistic and comprehensive. Key elements of the approach include:

*   preparing an inventory of watershed natural resources;
*   identifying the source of and degree of threats to those resources;
*   setting priorities for the protection and restoration of those natural resources.

The watershed planning process engages local stakeholders, scientists, advocacy groups, municipal officials, land-use decision-makers, and other civil groups. HREP staff and partners deliver watershed protection and restoration outreach tools to encourage implementation of water resource protection strategies. Thus education, communication, and partnerships between stakeholders provide the means to carry out watershed improvement projects. Synergies are sought in the use of tools and in the combination of measures implemented.

The assistance provided by the HREP to local initiatives includes: "Trees for Tribs" (for stream corridor re-vegetation), municipal stormwater and floodplain management advice, mapping, guidance for dam and other barrier removal, water-quality monitoring, land-use training, and grants. Specific actions carried out at local level include:

*   assessment of biological and chemical stream quality through stream bio-monitoring methods (e.g. tiered aquatic life uses);

- providing monitoring opportunities for volunteer and local leaders with encouragement to use results to inform land-use decision-making;
- assessment of motivations and incentives for protection of water resources on private land through reduction of impervious cover and other river stewardship measures;
- tracking of land-cover changes;
- provision of technical assistance and resources to active, locally led watershed groups on all significant tributaries of the Hudson to support partners in implementing water resource protection and restoration targets through intermunicipal watershed planning and implementation;
- integration of "green infrastructure" and stormwater retrofits into building regulation revisions to reduce pollution impacts from urban stormwater sources and enhance groundwater recharge;
- ensuring that all municipalities throughout the estuary watershed are aware of available water resource protection and restoration strategies;
- investigation of the feasibility of alternative wastewater options, such as decentralized approaches that promote groundwater recharge and water reuse.

### Action Agenda Goal 10

*Ensure that Hudson River water quality supports human benefits, including drinking water, swimming, fishing, navigation, and ecosystem protection*

High-quality river and estuary water benefits municipalities which draw their drinking water from the Hudson, swimmers, anglers, kayakers, riverfront parks, restaurants, marinas, and residential development. It also sustains the ecosystem of the river. Improvements achieved in water quality since the 1960s and 1970s have increased opportunities to sustain or expand all of these water uses. However, there are key challenges that remain to be overcome. Investment is needed to improve water and sewer infrastructure given ageing systems and growing demand, and to redesign stormwater systems to allow rainwater to replenish groundwater rather than flood streets and sewers. The HREP, working with New York City's Department of Environmental Conservation, is focusing on four primary strategies:

- seasonal disinfection of municipal wastewater discharges;
- reduction of combined sewer overflow impacts;
- local implementation and compliance with a Stormwater Permit Program to reduce runoff impacts;
- continued support for vessel waste pump-out facilities to maintain a "No Discharge Zone" status for the Hudson.

### Action Agenda Goal 8

*Promote public understanding of the Hudson River, including the life it supports, its role in the global ecosystem, and the challenges the river faces and how they can be met*

Since its inception the HREP has achieved a scientifically based understanding of the estuary and its watershed, but recognizes that the participation of citizens, river users, businesses, scientists, and community leaders is essential if this understanding is to be effectively applied in future management decisions. The knowledge base compiled by the HREP must be readily accessible to the public and local community leaders in formats that are understandable and "user friendly." Their use of this information to develop creative solutions to complex local problems may require technical assistance by the HREP or its partner organizations. The overall aim is to establish a citizenry and municipal admini-strations knowledgeable about the ecology and natural resources of the Hudson, and primed to support decisions that further improve land and water management.

To achieve the goal of promoting public understanding of the river and its watershed, the HREP implements specific actions under four main themes:

- creation of a network of places where the public can learn about the river and its watershed;
- provision of enhanced access to river information for the public;
- enhancement of school programmes for place-based river learning;
- improvement of the effectiveness of programmes serving educators, citizens, and administrators concerned about the river.

### Action Agenda Goal 12

*Track our progress and celebrate our successes*

This goal is recognized as vital for two reasons. First, the Hudson River Watershed is subject to dynamic and diverse processes of change. Within the landscape and its economy, patterns of development are changing in ways that can affect water quality and habitats. Changes in climate may also impact aquatic and terrestrial habitats, species distributions, and shoreline and coastal infrastructure. Meanwhile profound changes can already be observed in the river's ecosystem, for example, Zebra mussels have altered the food web with impacts on aquatic animal populations. Thus it is important to observe and record these changes to better understand and predict how they will affect the river. Second, the HREP considers it beneficial to reflect upon and celebrate its successes, evaluating what has been accomplished and looking toward future achievements. Fulfilling these two objectives motivates better-informed decisions and management actions, and the information generated provides the means to engage partners in effective action to minimize and mitigate harmful environmental impacts.

To this end the HREP has established working partnerships with local governments, business leaders, schools, and grassroots non-profit organizations. In further partnership with regional academic and research institutions the aim is to strengthen the scientific foundation of the Hudson River Estuary Program and to make it a model for scientific management through productive partnerships. As noted above, it is recognized that scientific information gathered by the programme's many studies needs to be interpreted into an accessible and understandable format to allow the public to monitor progress, evaluate the effectiveness of programme activities and participate in future decision-making that guides the programme. Work on an improved monitoring programme has included development of indicators, or "vital signs," that aim to measure the health of both human communities and natural systems within the watershed.

To achieve the goal of tracking progress and celebrating successes the HREP implements specific actions under four main themes:

- monitoring ecosystem condition and tracking programme performance;
- building partnerships (with the aim of engaging in implementation over 500 partners among municipalities, businesses, and non-profit organizations);
- building accessible databases;
- celebrating progress to promote the HREP and to bring heightened attention to the Hudson River and its many values.

Box 5.2 provides a simple but effective example of one of the means through which Action Agenda goals 8 and 12 are being achieved. It highlights both the role of information that is imaginatively presented to the public, and the use of cost-effective partnership working to achieve delivery.

---

**Box 5.2  The Atlantic sturgeon: the logo of the Hudson River Estuary**

The Hudson River Estuary logo depicts an Atlantic sturgeon, the Hudson's largest fish (see http://www.dec.ny.gov/lands/4920.html). This highlights the estuary's critical role as habitat for valuable fish and wildlife and the need to be vigilant in protecting this natural heritage. Through a partnership involving the NYSDEC, the New York State Department of Transportation, the New York State Thruway Authority, and the New York State Bridge Authority, the logo appears on signs by the bridges where major highways cross tributaries of the estuary. It reminds travellers that these streams are intimately connected to the mainstream, and that the health of the Hudson depends on the health of its watershed.

## Getting things done

### *Overarching governance arrangements and operating principles*

The Clean Water Act, formerly the Federal Water Pollution Control Act, is credited in large part with the reduction in pollution in the Hudson since 1972. This Act provided the means to improve municipal sewage treatment and to regulate point and non-point sources of water pollution. It has provided the essential legislative support to the relevant activities of the HREP.

However, the HREP has become much more than regulatory implementation by the NYSDEC. It has evolved to become an exceptional regional partnership for restoration, conservation, and protection of the Hudson River and its watershed. "Cornerstones" of its approach are that it is built on sound science and principles of ecosystem-based management. Supported by New York State, the HREP is hosted and coordinated within the NYSDEC. Hosted and coordinated are key descriptors because the programme functions in a highly cooperative manner through its partnerships with many organizations. These include local municipalities, non-government organizations, environmental groups, and research institutes. Annual funding of the HREP has been of the order of US $6 million, and these funds have been dispersed for implementation of the Action Agenda through the comprehensively sustained collaborative partnerships. For example, the HREP implements a significant portion of its priorities through contractual affiliations with academic and non-profit interstate partners.

The HREP is guided by the Hudson River Estuary Management Advisory Committee (HREMAC), which was established in 1987 under Section 11-0306 of the Environmental Conservation Law known as the Hudson River Estuary Management Act. The committee's volunteer members are appointed by the NYSDEC commissioner and include representatives of the commercial fishing industry, recreational anglers and other users, utility companies, local government, educators, researchers, conservationists, and marine trades and industry. Thus the committee represents interests directly involved in the estuary, and helps engage many representatives of the public in working collectively toward common goals. The committee meets three times per year and provides a forum to share the views and opinions of these diverse interests with state resource managers. The committee also serves to advise the NYSDEC on regulatory, policy, and other matters affecting the management, protection, and use of the Hudson River Estuary, its tributaries and shorelands, including implementation, evaluation, communication to stakeholders, and the public sharing of the results of the HREP (NYSDEC, 2013).

### *Watershed management programmes*

As noted above, of most relevance to the focus of this book is the development of sub-basin or tributary protection programmes under Goal 4 of the HREP's

Action Agenda. A primary governance instrument by which such watershed management may be achieved is through intermunicipal agreements. Such agreements are provided for under New York State law, giving authority for towns, cities, or villages to enter into agreements to foster cooperation or integrated planning and regulation where the issues transcend individual municipalities. The challenges of establishing and implementing such agreements are discussed further below.

Under the relevant state law municipalities are enabled to work cooperatively to carry out comprehensive planning and land-use regulation at sub-basin scale to protect water quality in streams and drinking water, maintain stream flows and water availability, conserve floodplains and stream corridors, and minimize flood impacts. To achieve these objectives the HREP also works closely with other groups that focus on tributary and watershed management and which can operate across sub-basins and at watershed scale. These groups include the Scenic Hudson, *Clearwater*, Riverkeeper, and the Hudson River Watershed Alliance (HRWA). All are strong regional activist groups that work collectively to safeguard the Hudson and protect its natural and cultural resources.

The HRWA has been particularly significant for HREP in this context. Its mission is to protect, conserve, and restore the water resources of the Hudson River Basin through collaborative outreach, education, networking, science, information-sharing, and technical assistance by and for the stakeholders of the region (HRWA, 2013). The HRWA is a network of groups and individuals concerned about, and working on, protection of groundwater, lakes, ponds, wetlands, reservoirs, tributaries, and river and estuary itself in the Hudson watershed. It is composed of member organizations and individuals that subscribe to the HRWA's vision and approach, and other interested parties recognized as Participating Organizations. A board of directors representative of sectors and geographic locations meets at least quarterly to coordinate and direct HRWA activities. The HRWA has the status of a non-profit public charity and its activities are funded by grants, contracts, and contributions from member organizations and individuals.

HRWA members comprise a wide range of governmental agencies, research and educational institutions, environmental organizations, and grassroots groups. A central purpose is to create common ground among these diverse groups, and to avoid polarization. Thus the HRWA seeks to facilitate dialogue and cooperation, but does not participate in, or take positions on, site-specific issues. This has made it a useful partner for the HREP which has been able to rely on the HRWA for its assistance in networking and reaching out effectively to a wide range of organizations and interest groups. The neutral stance and standing of the HRWA, and the widespread trust that this engenders, have also been of particular value in helping to develop interjurisdictional cooperation for HREP actions.

The HREP's staff members provide information to connect local governments to the natural resources in their communities and about the principles and legal framework for sound land-use planning. This assistance, along with funding support through the Estuary Grant Program, allows counties, towns, and villages in the Hudson River Valley to take ownership of their resources and define the future of their communities while contributing to the overall health and beauty of the region. For example, using advanced technology the river is being mapped to help communities manage river habitats and human uses. The HREP has also expanded opportunities for citizens to connect to the river by upgrading boat launches and preserving valuable open space in sight of the river. It also assists in cleaning up water pollution from persistent chemicals and human waste. Funded projects that address the HREP's goals include helping to manage and restore key species such as striped bass and bald eagles, protect key habitats (e.g. underwater grass beds and tidal wetlands), and conserve plants, animals, and habitats in the river basin landscape.

### Challenges in promoting watershed management within the Hudson Valley

A major challenge for the HREP in promoting watershed or catchment planning and management in the Hudson Valley is conflict between "top-down" management and grassroots preference for self-governance. Watershed management generally represents a midway point between these two frequently conflicting levels of government as a form of regionalism. Cullingworth (1993, p. 199) describes regional planning in the United States as a "succession of false starts and disappointed hopes." Such false hopes especially apply to watersheds. Ingram (1973) suggests a principal reason for this disappointment is that regional institutions are prescribed on technical and economic terms, whereas the over-riding challenge in watershed management is that cross-jurisdictional boundaries are more likely to be of a political, institutional, and legal character. Municipal jurisdictions fiercely defend their prerogatives, and this is especially the case in the lower Hudson Valley. Referring to this governmental "balkanization" Wood and Almendinger (1961, p. 1) stated that in this region of New York State, "explorers of political affairs can observe one of the great unnatural wonders of the world: that is, a governmental arrangement perhaps more complicated than any other that mankind has yet contrived or allowed to happen." More recently Benjamin and Nathan also studied what they term the "crazy quilt of local governments" (Benjamin and Nathan, 2001, p. 3). Regardless, watershed planning and governance must accommodate the reality of the political fiefdoms highly protective of their autonomy that make up this "crazy quilt." Hence watershed planning necessitates building upon the interests and functions of the local governments upon which such planning must ultimately depend.

Governance of water resources linked to land use in particular has a marked local character in the Hudson Valley. Local governments there strongly defend their right to make their own land-use decisions. Under the home-rule provisions of the New York State Constitution the local municipalities enjoy autonomy with respect to "purely local matters" (Department of State, 2010). This autonomy is fundamental to any discussion of regional governance.

Municipal Home Rule Law and the Statute of Local Governments give local governments rights to enact local laws concerning local affairs, provided the laws do not conflict with state laws (Krane *et al.*, 2001). Water supplies are generally a responsibility of local governments, and there is a possessive pride of ownership in these water systems.

Home-rule authority naturally does not extend to managing water resources on a watershed scale encompassing multiple jurisdictions. Apart from the lack of jurisdictional authority, local governments generally lack the capacity, capital, and expertise needed for watershed management. Although the state may possess the capacity, deference to home rule prevails. However, as demonstrated by the Hudson River Estuary Program, when home rule prerogatives are respected, the state can play a highly significant role in assisting local municipalities by providing technical and financial resources (Krane *et al.*, 2001). Thus watershed planning and management need not be an invasion of home-rule authority. Political accountability and public participation, the core of home-rule prerogatives, can be institutionalized in cooperative watershed management. The challenge is to overcome the patchwork quilt of independent local governments that otherwise confounds cooperation.

Obviously, catchment management, which typically crosses jurisdictional lines, does not immediately fall under the home-rule powers encompassing only issues "of purely local concern" (Department of State, 2010). However, such management may be accomplished through intermunicipal cooperation. Intermunicipal cooperation is encouraged under New York State law by providing municipalities with the authority to enter into intermunicipal agreements (IMAs) (McKinney, 2008). IMAs are "cooperative or contractual arrangement[s] between two or more municipalities." New York State specifically authorizes municipalities and counties to create regional planning boards and joint-purpose municipal corporations. An example established under this authority is the Hudson Valley Regional Council, comprising the counties: Dutchess, Orange, Putnam, Rockland, Sullivan, Ulster, and Westchester (Hudson Valley Regional Council, 2013). IMAs offer a flexible means that enable municipalities to protect water resources, extending across jurisdictions that could otherwise be threatened by disconnected balkanization of land-use decisions.

### The example of the Moodna Creek Watershed

The Hudson River Estuary Program strongly promotes IMAs as an instrument to foster tributary watershed management. A current foremost example is the Moodna Creek Watershed in the south-eastern corner of the Hudson River greater watershed. The Orange County Water Authority and County Department of Planning, assisted by a working group of local stakeholders, and with financial support from the Hudson River Estuary Program, created the Moodna Creek Watershed Conservation and Management Plan. The plan was finalized in March 2010 (Orange County Water Authority, 2013a). It is designed to address comprehensively local and intermunicipal objectives and to support source water protection, regional watershed planning, and research and monitoring of water resources. The goals of the Moodna Creek Watershed Conservation and Management Plan are to:

- summarize existing conditions in the watershed;
- identify and describe issues that are important to local communities and stakeholders, including assets, existing problems, risks, and opportunities;
- develop a prioritized list of action items and recommendations for addressing identified issues.

The plan covers the major topics of: land use, water quality, wastewater treatment, water supplies, climate and water resources, flooding issues, ecological resources, riparian corridors, and recreation. Planning and goals for the watershed plan include facilitating and developing a coordinated intermunicipal programme. A highest priority recommendation of the plan was the creation of the Moodna Creek Watershed Intermunicipal Council to cooperatively implement the recommendations of the plan and additional watershed issues as they may arise. The Moodna Creek Watershed area overlaps the jurisdictions of 22 municipalities in Orange County, although for several the overlapping area is very small. On October 22, 2010, ten towns, four villages, and Orange County itself, signed an intermunicipal Memorandum of Agreement establishing the Council (Orange County Water Authority, 2013b). The primary purpose of the Council is to "continue to work together across municipal boundaries in order to protect, conserve, and enhance the water resources of the Moodna Creek and its watershed." Specific functions of the Council are to:

- secure and share public and private grants available to address issues pertaining to watershed protection and management;
- utilize each undersigned party's ability to address issues pertaining to the Creek and its watershed;
- create an avenue for intermunicipal dialogue for addressing water quality and water quantity issues;
- consider implementing the Moodna Creek Watershed Conservation and Management Plan;
- develop educational programmes on watershed planning, flooding, pollution prevention, stormwater management, biological resources, and other best management practices for individuals and municipalities;
- coordinate other organizational efforts in each municipality that impact or benefit the resources of the Moodna Creek Watershed; and
- benefit watershed municipalities, individually and collectively, by integrating protection of watershed resources with economic and social policies.

The Moodna Creek Watershed Intermunicipal Council meets quarterly. Members of the public are welcome to attend the meetings. The Council operates on the basis of goodwill and shared purposes between its parties. It also has strong institutional support from Orange County, through the Orange County Water Authority, and from the New York State Hudson River Estuary Program. In terms of its institutional formation and functions, the Moodna Creek Watershed Program serves as an outstanding watershed governance model for other watershed areas in the Hudson Valley.

### Monitoring strategy

A key element of the monitoring strategy is the Hudson River Environmental Conditions Observing System (HRECOS), a network of real-time monitoring stations on the estuary. The stations are geographically distributed from Schodack Island to New York Harbor and monitor river conditions every 15 minutes. The HRECOS provides another example of a collaborative approach as it is operated by a consortium of partners from government and the research community. Partners include the NYSDEC, the Hudson River Foundation, the National Estuarine Research Reserve, the US Geological Survey, the Cary Institute of Ecosystem Studies, the Lamont Doherty Earth Observatory, and the Stevens Institute of Technology. Funding is similarly based on partnership working with contributions from the United States Environmental Protection Agency, National Oceanic and Atmospheric Administration, Hudson River Foundation and the HREP. Progress and achievements by the HREP are celebrated with all partners though annual National Estuaries Days, publication of stories about what has been learned, and the making of information accessible to the public on the web and through other media.

### Outcomes

The HREP and its Action Agendas emerged as a response to both poor river conditions in the 1960s and 1970s, and the protests of activist conservation groups. Four Action Agendas have been issued and implemented by the HREP since 1996. Built on sound science and extensive public input these plans set goals and targets for progress with the intention that these should be objectively measured and publicly reported. They also provided a framework to guide local governments, state agencies, and non-profit organizations in working collaboratively to common purpose. As a result it has been possible for the HREP and its partners to make coordinated investments in a wide range of watershed management and improvement interventions including river access, pollution reduction and clean-up, open-space conservation, habitat conservation and fisheries management, and land-use planning and regulation.

The accomplishments and outcomes of the HREP since 1996 are numerous and diverse and can only be briefly summarized here. With regard to water quality the programme's target to achieve swimmable water in the Hudson has been significantly advanced through state grants (and leveraged matching funds) to municipalities to assist them in meeting water quality standards. Measures taken include seasonal disinfection of wastewater discharges and controls for combined sewer overflows. Contaminant reduction and clean-up projects have also been implemented with harbor authorities and industrial partners, although these still face the challenge of a legacy of industrial pollutants such as PCBs (polychlorinated biphenyls) present in sediments many years after use has been banned. In 2003 the entire Hudson Estuary was designated a "No Discharge Zone" to prohibit the discharge of all waste from vessels, while provision was made for 15 pump-out stations along the river for use by recreational boaters. Since 2008 real-time, internet-accessible monitoring by HRECOS has provided the means to assess outcomes and to understand the responses of the river to changing conditions (NYSDEC, 2010).

The HREP has also raised local awareness of how the Hudson Valley's rivers and streams relate to the estuary ecosystem and related conservation concerns. As partners the Hudson Basin River Watch (HBRW, 2013) has connected schools and adults to the river through its citizen water-quality monitoring programme, and the HRWA has provided a forum for regular public education and outreach programmes. As a result, community-based watershed conservation planning and implementation is ongoing in 15 tributary watersheds, supporting the development of a river-stewardship ethic throughout the valley. The HREP has partnered with county and local governments to identify and adopt strategies for protecting water resources, such as the inclusion of better site-design principles in zoning, riparian and wetland buffer protection ordinances, and local stormwater laws. By 2009, the HREP's "Trees for Tribs" initiative had protected more than 32,000 feet of stream buffers at more than 70 sites, and over 10,000 native trees, shrubs, and grasses had been planted by 1,200 volunteers to protect water quality and stream habitat (NYSDEC, 2011).

Key habitats in the river have been mapped, including the estuary's tidal wetlands, submerged aquatic vegetation beds, more than two-thirds of the estuary's bottom and all of its shoreline. An Estuary Training Program was established and now provides high-quality capacity development for habitats, resource management, technical skills and process skills for local decision-makers, community leaders, environmental groups, local land trusts, natural resource managers, and regulators. Stock condition assessments and updating amendments to interstate fishery management plans through the HREP have enabled the NYSDEC to manage fish populations which drive annual recreation and tourism expenditures of US $7.5 million (NYSDEC, 2010). NYSDEC (2010) also reports that through the HREP more than 5,000 community leaders have received training and information resources to support achievement of conservation locally.

The first-ever wildlife and habitat monitoring plan for the Hudson Valley was developed to better understand and report the status of important biological resources to state and local leaders. Mapping of the Hudson Valley's biodiversity resources in partnership with the Natural Heritage Program documented for the first time the global, regional, and statewide significance of the area's plants and animals. The HREP has also provided technical assistance and funded nearly 120 projects to develop a network of more than 20 environmental education sites along the river. A curriculum developed by the HREP is being used in classrooms in 40 Hudson Valley schools and other educational initiatives.

Concerned to meet the challenges of climate change, the HREP has established a network in the region that includes more than 100 community leaders. Models under development from this collaboration will assist in predicting rises in sea level in the Hudson, and provide vulnerable communities in the tidal portion of the Hudson with up-to-date information. A programme for "Climate Smart Communities" aims to assist municipalities in reducing their carbon emissions and in becoming more energy efficient (NYSDEC, 2011).

## Concluding reflections

Since 1996 the HREP has invested in the development of partnerships with state and federal agencies, local governments, and other involved groups and organizations, intensive applied research, adaptive approaches to management and public outreach. These efforts have been for the betterment of the Hudson River Estuary and of the land and communities of its surrounding watershed. Both work and monitoring are ongoing but significant achievements have been made in environmental conservation and in developing the capacity of both state agencies and local communities to become stewards of natural resources. The activities of the HREP have helped to restore and conserve the natural environment of the Hudson Watershed, and they have contributed stimulus to the regional economy and enhanced the Hudson Valley as a destination for tourism (NYSDEC, 2011).

In a programme of this reach and diversity, interorganizational partnerships at many levels provide the means to combine complementary expertise and resources, coordinate actions, and leverage additional funding. Public participation can guide programme formulation, prioritization and decision-making, while public outreach and education builds support, assists funding, and increases impact through voluntary action and compliance.

Key lessons include learning that management of the watershed "must simultaneously occur at the scale of each watershed's tributary streams and on the scale of the Hudson River itself" (Cuppett and Urban-Mead, 2010, p. 11). Each watershed plan should include development objectives, priorities for implementation, and list individuals responsible for their implementation. At all levels—regionally, watershed, county, city, town, and village—formal agreements to work together need to be explored and developed. Given the scale of the Hudson River Watershed and the complexity of its challenges, intermunicipal cooperation based on agreements and shared objectives are essential.

A continual challenge for the HREP is measuring its success. As with many government programmes evaluations can become based on process measures such as the number of people trained, workshops offered, or trees planted. This is not meaningless in this context because these outputs are essential to the achievement of environmental conservation and protection objectives. Yet the HREP and its partners are striving to make monitoring of its activities more meaningful in terms of the natural resources being protected or restored, and the change in behavior of target audiences. Early indications suggest that there is a recognizable connection between programme education and outreach and behavioral change, such as municipalities adopting conservation practices, policies, and ordinances. As a result, this type of evaluation will continue to be refined and adapted (Cuppett and Urban-Mead, 2010).

## Note

1    Further geographical description and maps for the Hudson River Estuary and its watershed are available from the websites of the NYSDEC (http://www.dec.ny.gov/lands/26561.html) and the HREP (http://www.dec.ny.gov/lands/4920.html).

## References

Benjamin, G. and Nathan, R. P. (2001) *Regionalism and Realism: A Study of Governments in the New York Metropolitan Area*, Brookings Institution Press, Washington, DC.

Benke, A. C. and Cushing, C. E. (2005) *Rivers of North America*, Academic Press, Elsevier, San Diego, CA.

Cullingworth, J. B. (1993) *The Political Culture of Planning: American Land Use Planning in Comparative Perspective*, Routledge, New York.

Cuppett, S. and Urban-Mead, R. (2010) *Hudson Valley Water: Opportunities and Challenges*, CRREO Discussion Brief 4: Center for Research, Regional Education and Outreach, State University of New York at New Paltz, New Paltz, NY.

Department of State (2010) The Constitution of the State of New York, Department of State Division of Administrative Rules, Albany, NY. www.dos.ny.gov/info/constitution.htm, accessed August 19, 2013.

Dunwell, F. F. (2008) *The Hudson: America's River*, Columbia University Press, New York.

Foderaro, L.W. (2007) 'In Westchester, cleaning up after one flood while planning for the next', *New York Times*, November 13.

Gordon, D. and Kennedy Jr, R. F. (1991) *The Legend of City Water: Recommendations for Rescuing the New York City Water Supply*, The Hudson Riverkeeper Fund.

HBRW (2013) *Hudson Basin River Watch*, www.hudsonbasin.org/, accessed August 19, 2013.

HRSC (2013) *Hudson River Sloop Clearwater*, About Clearwater, http://www.clearwater.org/about/, accessed August 16, 2013.

HRWA (2013) *Hudson River Watershed Alliance*, www.hudsonwatershed.org, accessed August 19, 2013.

Hudson Valley Regional Council (2013) *About Us*, www.hudsonvalleyregionalcouncil.org/about-us/, accessed August 19, 2013.

Ingram, H. M. (1973) 'The political economy of regional water institutions', *American Journal of Agricultural Economics*, vol. 55, no. 1, pp. 10–18.

Krane, D., Rigos, P. N. and Hill, M. B. (2001) *Home Rule in America: A Fifty State Handbook*, CQ Press, Washington, DC.

McKinney, W. M. (ed.) (2008) *McKinney's Consolidated Laws of New York Annotated: With Annotations from State and Federal Courts and State Agencies*, West Publishing Company, St Paul, MN.

Moira Productions (2006) *Swim for the River: Directory of Hudson River Environmental Organizations*, http://www.swimfortheriver.com/pdf/directory.pdf, accessed August 16, 2013.

NYSDEC (2010) *Hudson River Estuary Action Agenda 2010–2014*, New York State Department of Environmental Conservation, Albany, NY.

NYSDEC (2011) *Hudson River Estuary Program Report of 15 Years of Progress*, New York State Department of Environmental Conservation, Albany, NY.

NYSDEC (2012) *Hudson River Estuary Program*, New York State Department of Environmental Conservation, Albany, NY.

NYSDEC (2013) *Hudson River Estuary Management Advisory Committee*, New York State Department of Environmental Conservation, www.dec.ny.gov/about/46924.html, accessed August 19, 2013.

NYSDEC (2014) *About DEC*, New York State Department of Environmental Conservation, www.dec.ny.gov/24.html, accessed December 16, 2014.

Orange County Water Authority (2013a) *Moodna Creek Watershed*, www.waterauthority.orangecountygov.com/moodna.html, accessed August 19, 2013.

Orange County Water Authority (2013b) *Moodna Creek Watershed Intermunicipal Council*, www.waterauthority.orangecountygov.com/moodna_council.html, accessed August 19, 2013.

Padavan, F. (1988) *Muddling Through: A Legislative Review Examining the Ability of the New York City Water System to Meet the Demand of Southeastern New York in the 21st Century with Recommendations*, New York State Legislature, Senate Standing Committee on Cities and City of New York, Albany, NY.

Riverkeeper (2013) *About us*, www.riverkeeper.org/about-us/, accessed August 16, 2013.

Rosenzweig, C. and Solecki, W. (2001) 'Climate change and a global city: Learning from New York', *Environment*, vol. 43, no. 3, pp. 8–18.

Scenic Hudson (2013) *About Scenic Hudson*, http://www.scenichudson.org/about, accessed August 16, 2013.

Wood, R. C. and Almendinger, V. V. (1961) *1400 Governments: The Political Economy of the New York Metropolitan Region*, Harvard University Press, Cambridge, MA.

# 6 Healthy Waterways, South East Queensland, Australia

*Laurence Smith, David Benson and Diane Tarte*

## Introduction

This chapter presents the lessons from Healthy Waterways (HW), a unique organisation that brings people together across multiple sectors to manage waterways in South East Queensland (SEQ), Australia. HW is an independent, not-for-profit, non-governmental, membership-based organisation that works to protect and improve waterway health. It does so in a region where rivers, lakes and the coast are recognised to be an integral part of lifestyles and the economy. HW facilitates planning and coordination of efforts at local and regional levels by its network of member organisations. Collectively these aim to deliver a shared vision for waterway improvement and health.

As elsewhere in the driest developed continent on Earth, water abstraction and allocation are a pre-occupation for policymakers and catchment managers but SEQ also experiences extreme flood events. Catchments have become degraded by the expansion of 'broadacre' farming (large-scale crop operations including oilseeds, cereals, pulses and sugar cane), intensive livestock farming and horticulture, coastal urbanisation, industrialisation and other infrastructure. Water quality thus proves to be a difficult management issue, with problems varying according to geographical context. Acid sulphate soil leaching has impacted estuarine and coastal water quality in some areas, while in most of Queensland eutrophication and sedimentation of the Great Barrier Reef and other coastal zones is primarily caused by nutrient-rich agricultural runoff (McDonald and Roberts, 2006; Great Barrier Reef Marine Park Authority, 2009).

Established from research conducted in the early 1990s, HW provides leadership and enhanced collaboration for management of water quality in SEQ, and works closely with multiple governmental and non-governmental actors, including state environment and natural resource management (NRM) agencies, local government, industry, researchers, community organisations and the wider public. Development of the organisation provides lessons on ways to adaptively manage problems at the catchment scale by creating broad-based awareness, building partnerships and engaging action by multiple actors at different scales and locations. As a case study it illustrates how regional collaboration across the whole water cycle can be key to achieving clean water and a healthy environment.

## Context and drivers

### Catchment management in Australia

Catchment management in Australia differs in distinct ways from practice in Europe and North America. The main difference arises in the emphasis given to holistic NRM at the catchment scale. Rather than narrowly focusing on the problem of water quality alone, Australian approaches have attempted to apply and develop the concept of integrated catchment management (ICM). Since the 1980s this has evolved through both 'bottom-up' local responses and 'top-down' state and Commonwealth (national government) initiatives.

Initially many NRM problems in Australia were tackled by local groups that built and drew upon community support. Then in the 1980s, policymakers at the state and Commonwealth levels began to recognise a range of critical land-use problems that ranged from soil salinisation and erosion to loss of biodiversity, and to prioritise catchment-scale responses that required collaboration by stakeholders including community-based groups. A key step in this process was national consultation undertaken by the Commonwealth government in the late 1990s, resulting in publication of recommendations for NRM policy reform (Commonwealth of Australia, 1999). Successive Commonwealth funding initiatives have also sought to promote collaborative approaches for NRM, beginning with the Decade of Landcare in the late 1980s, the Natural Heritage Trust (NHT) policy from the mid-1990s and Caring for Our Country (CfOC) since 2008.

State governments have also encouraged ICM, their policies usually becoming aligned to integrate local programmes with regional-scale institutions and qualify for Commonwealth funding, initially under the NHT and now CfOC, but with some variation in approach by state. In New South Wales (NSW), thirteen regional-scale catchment management authorities (CMAs) were established in 2003. With direction from a statutory board incorporating government actors and community representatives CMAs are responsible for developing Catchment Action Plans (CAPs) and facilitating NRM investment in their regions by utilising state and Commonwealth funds (NSW Government, 2013). In Victoria, NRM has been similarly devolved to ten CMAs, each implementing an integrated Regional Catchment Strategy (Park and Alexander, 2005), and South Australia (SA) oversees eight regional NRM Boards and associated local groups via a state NRM Council. Western Australia has six regional NRM bodies also overseen by a state NRM Council (Farrelly and Conacher, 2007), Tasmania has three and Northern Territory one.

In contrast, Queensland has taken a less interventionist approach than other states. Its NRM policy framework supports pre-existing non-governmental catchment organisations that have gained Commonwealth and state agency funding, and aims to 'integrate and align NRM effort at the landscape level, define how priorities are set for Queensland Government regional NRM investment, and identify opportunities for improving NRM arrangements'

(Department of Environment and Resource Management, 2011, p. 6). The state seeks to provide flexible support and coordination for existing not-for-profit, community-governed regional organisations compared to the more directive leadership potentially provided by the recently established statutory-based CMAs in NSW, Victoria and SA.

## Location and key characteristics of the catchments

Operating within the policy framework just described, HW facilitates collaborative waterways management across SEQ, a region situated within Australia's largest and most bio-physically diverse state. The region includes nineteen major catchments and many sub-catchments centred on the Brisbane River and Moreton Bay[1] that have a total area of 21,672 square kilometres.

Within these catchments are many different types of waterways, ranging from streams arising from the Great Dividing Range of mountains to the west to the shallow coastal zone of Moreton Bay itself in the east. The bay is a major recreation and tourism destination and contains protected wildlife areas that support many protected species including dugongs, migratory birds, sea turtles and fish. The region's waterways also sustain a number of other economic activities, contribute to its overall quality of life, and are recognised as significantly defining 'the character of South East Queensland' (SEQ HWP, 2007).

Critically for water resource management the region contains several large urban areas and has a rapidly expanding population. The largest urban centre is Brisbane with some one million residents, now Australia's third largest city by population. Advancing rapidly towards Brisbane from the south is the Gold Coast conurbation centred on the towns of Tweed Heads, Coolangatta and Surfers Paradise. Formerly a string of small tourist resorts, the Gold Coast is now merging into the fastest-growing urban area in the country, attracting inward migration from southern Australian states and abroad. While the total population of the SEQ region currently stands at 2.8 million, current predictions estimate that it will expand to over four million by 2026.

Although its economy is becoming increasingly diverse, with growing service, manufacturing and education sectors, tourism and primary industries including agriculture remain very important. The region's coastline and waterways are a major draw for visitors and together nature-based tourism, local recreation and recreational fishing contribute over  AUS $3,700 million per year to the regional economy and 7 per cent of employment (SEQ HWP, 2007). In comparison, primary industries including agriculture, commercial fisheries and aquaculture generate approximately AUS $1,400 million per year.

## Specific water quality and quantity problems and their severity

Settlement patterns and economic development have generated serious water quality problems, manifest primarily in high loads of nitrogen, phosphorus and sediments being exported from the catchments into the receiving coastal waters. These three pollutants have three major sources: point source, urban diffuse and rural diffuse. The Brisbane River, its tributaries and Moreton Bay have been contaminated by point sources of pollution from heavy industries and wastewater treatment plants. Increasing point source loads are driven mainly by population and economic growth. Diffuse or non-point sources also present a significant challenge to water quality as sediments and nutrients (nitrogen and phosphorus) emanating from urban areas, agriculture and land clearance are washed into the rivers and Moreton Bay by stormwater runoff across the region's catchments.

Predicted increases in diffuse pollutant loads result primarily from the rapid urbanisation of the region which causes accelerated clearance of natural vegetation and increases surface water runoff and associated diffuse pollution. Compared to estimated pollutant loads in 2004, under a 'business as usual' scenario with no improvements in management, nitrogen loads were predicted to increase by 49 per cent (point sources), 55 per cent (diffuse urban sources) and 4 per cent (diffuse rural sources); phosphorus loads by 39 per cent (point), 67 per cent (diffuse urban) and 5 per cent (diffuse rural); and sediment loads by 63 per cent (diffuse urban) and 5 per cent (diffuse rural). This illustrates the dominance of increasing point source and diffuse urban water pollution with rapid urbanisation (SEQ HWP, 2007).

One impact of pollution on the waterways is blooms of both freshwater and marine cyanobacteria, which are generally referred to as nuisance algal blooms (Box 6.1). Temporal and spatial variability of rainfall increases the risks to Moreton Bay from land-based pollution. Highly seasonal, 'flashy' flow events exacerbate sediment and nutrient runoff, thereby complicating catchment and water management. Climate change scenarios indicate that such events will increase, although overall rainfall is likely to decrease.

---

**Box 6.1 Cyanobacteria growth**

HW coordinated a major research programme into the marine blooms primarily caused by the cyanobacteria, *Lyngbya*. This work identified that nitrogen is a key limiting element in *Lyngbya* growth in Moreton Bay and that bio-available iron in runoff from the catchment also mobilises *Lyngbya* blooms. Effects from these cyanobacteria include toxicity to humans and impacts to ecological resources including smothering of seagrass beds.

---

Population growth is also placing additional strain on the water supply capacity of catchments with demand increasing for potable water and additional resources for domestic, industrial and recreational uses; this was exacerbated by significantly increased per capita urban water usage levels during the 1990s and 2000s. The severe drought between 2000 and 2007 further exacerbated these pressures and resulted in a 'water crisis' in the region in 2007 with water supply levels falling to 15 per cent of water storage capacity. Consequently severe water use restrictions were introduced and AUS $7 billion worth of infrastructure was built to 'drought proof' the region. This included four advanced water treatment plants for wastewater plant discharges, a desalinisation plant and the establishment of a water grid across the region. The changing patterns of rainfall predicted as a result of climate change will further complicate the future management of water resources.

Combined, these factors present significant challenges and raise the question of whether the economic development of SEQ can be sustained while maintaining healthy aquatic ecosystems. According to HW the risks of a 'business-as-usual' approach for ecosystems, the local economy and lifestyles are high (SEQ HWP, 2007). As a result of unchecked pressures on local ecosystems HW predicts declining water quality from the predicted increases in sediments and nutrient loads. Consequent negative impacts are likely to include higher water treatment costs, risks for the region's tourism industry, damage to commercial fisheries, aquaculture and recreational fishing, declines in agricultural production, increasing human health risks and biodiversity loss (SEQ HWP, 2007). The region's waterways provide valuable ecosystem services that sustain increasing tourism, high waterfront property values, better public health, a good quality of life and cheap and reliable water supplies for farmers (SEQ HWP, 2007). Thus sustainable water management that balances the needs of competing water uses has become a key aim.

### Key events and developments

HW has its origins in attempts to counter chronic pollution of the Brisbane River. Established by the British as a penal colony in the 1820s, the expanding city of Brisbane became the capital of the newly independent state of Queensland in 1859. At this time, the waters of the Brisbane River were naturally clear and contained many fish species. After 1862, however, dredging of the river bed was conducted to both improve navigation and extract sand and gravel. This extraction combined with vegetation clearance in the catchment increased the sediment load in the river water, reducing visibility and giving it a muddy appearance. Dredging continued well into the twentieth century, yet contemporary accounts of the river from the 1950s suggest it was still clean enough for recreational swimming. This situation changed significantly in the 1960s as the population continued to increase, underpinned by industrial and urban development. Wastewater treatment works built downstream of the city to service its expanding population discharged primary-treated effluent directly into the river. A number of large food-processing plants and abattoirs discharged untreated effluent directly to both the local streams and the Brisbane River. By the early 1970s, the Brisbane River was a health hazard, finally prompting reluctant state politicians to intervene. Queensland introduced a Clean Water Act in the mid-1970s but it was poorly enforced, resulting in local community activists initiating a campaign to clean up the river (Gregory, 1996).

HW (initially called the Moreton Bay Waterways and Catchment Partnership, and then SEQ Health Waterways Partnership) has its roots in this period. Microbiologist Di Tarte and other members of the Queensland Littoral Society (QLS) started conducting water quality tests of the Brisbane River which revealed significant levels of pollution. Yet local and state governments of the time remained largely unconcerned about the threat to the environment. Little action was taken for over a decade. As political attitudes towards the environment began to change throughout Australia in the late 1980s and early 1990s, demands to clean up the river grew. A book published by the Australia Littoral Society (formerly QLS) in 1990 described the pollution impacts (Davie *et al.*, 1990). In response, Brisbane City Council and its Lord Mayor Sallyanne Atkinson took the initiative by introducing the idea of reinventing Brisbane as the 'river city'; a concept driven forward by her successor, Jim Soorley, in the mid-1990s. From this emerging context, the Brisbane River Management Group (BRMG) and Policy Council were formed by the Queensland Government Department of Environment in 1993 and the Brisbane River and Moreton Bay Wastewater Management Study (the 'SEQ Study') was established by six local governments and the Queensland Government Department of Environment in 1994. These two initiatives combined in 2001 to form the basis of the current partnership.

## Getting started: a process of evolution and partnership building

Development of HW and its programmes from these beginnings involved a process of partnership-building over two decades, during which the BRMG and 'SEQ Study' evolved into more institutionalised forms (Figure 6.1). Each stage of development was evaluated to assess progress and formulate future plans.

From 1994 to 1997 (Figure 6.1) the 'SEQ Study' undertook in-depth scoping to review relevant literature, bring together engineering and scientific expertise, and establish priorities for subsequent study. A Scientific Advisory Group (now the HW Scientific Expert Panel) was formed from representatives of local universities, state and Commonwealth agencies, water companies and local councils to peer review research progress and ensure high standards. During this period a hydrodynamic transport model was developed and significant issues identified for further research. Consultative processes were conducted via a Key Stakeholder Advisory Group and stakeholder workshops focused on engaging farmers, local governments and the community (Figure 6.2).

From 1997 to 1999 (Figure 6.1) activity focused on Moreton Bay and its estuaries and the impacts of point source pollution. A primary objective was to develop a Water Quality Management Strategy and a set of necessary tasks was identified with stakeholders and the Scientific Advisory Group, aiming to generate the data required for development of a receiving water quality model

| 1994–1997 | 1997–1999 | 1999–2002 | 2002–2005 | 2005–2008 | 2009 onwards |
|---|---|---|---|---|---|
| **Stage 1** | **Stage 2** | **Stage 3** | **Integrated partnership** | **Integrated partnership** | **Integrated partnership** |
| **Scoping** | **Bay and estuaries focus** | **Rivers and catchment focus** | **Waterways management land to sea; sustainable catchment** | **Healthy catchments: healthy waterways** | **Whole-of-water cycle management; healthy country and water sensitive cities** |
| • Background studies | • Pilot studies<br>• Design of estuarine and marine monitoring programme | • Ongoing studies<br>• Estuarine and marine monitoring programme<br>• Design of freshwater monitoring programme<br>• Regional decision support tools<br>• Regional process studies | • Integrated freshwater, estuarine and marine monitoring programmes<br>• Design of load-based and human health monitoring programme<br>• Decision support tools for implementation<br>• Process studies<br>• Socio-economic studies<br>• Sustainable loads | • Integrated ambient (freshwater, estuarine marine) and event-based monitoring programmes<br>• Decision support tools used for management scenarios<br>• Process studies underpinning strategy development<br>• Sustainable loads | • Integrated waterways health accounting framework<br>• Management strategy evaluation framework (environmental, social, economic, institutional)<br>• Integrated decision support tools<br>• Interactive data integration system |
| • Six local councils<br>• State and federal governments | • Six local councils<br>• State and federal governments | • Nineteen local councils<br>• State and federal governments | • Nineteen local councils<br>• State and federal governments<br>• NRM initiatives | • Nineteen local councils<br>• State and federal governments<br>• NRM initiatives<br>• Water initiatives | • Ten local councils<br>• State and federal governments<br>• NRM initiatives<br>• Water initiatives |

*Figure 6.1* Key stages of development in the Healthy Waterways partnership and management strategy (Source: D. Tarte, 2010)

| | 1994–2000 | 2001–2007 | 2008–2010 |
|---|---|---|---|
| **Policy** | • Brisbane River Policy Council<br>• Brisbane River and Moreton Bay Wastewater Management Strategy Strategic Advisory Group | • Regional Coordination Committee<br>• SEQ HW Partnership Policy Council | • Regional Coordination Committee<br>• CEO Natural Resource Management Committee |
| **Scientific and technical** | • Scientific Advisory Group<br>• Technical Advisory Group | • Scientific Expert Panel (SEP)<br>• Advisory Groups of SEP (4)<br>• Technical Advisory Group | • Scientific Expert Panel (SEP)<br>• Advisory Groups of SEP (4)<br>• Technical Advisory Group |
| **Consultative** | • Key Stakeholder Advisory Group<br>• Workshops around the regions with relevant stakeholders from community, local governments and farmers | • CEO Committee<br>• Regional Implementation Groups (4)<br>• Community Industry Advisory Group<br>• Various SEQ HWP Programme Steering Committees (4)<br>• Science 'road shows', workshops and public meetings around the region | • SE Regional Coordination Group<br>• Various SEQ HWP Programme Steering Committees (4) |
| **Funding** | Commonwealth, state and local government's – project based | Regional programme: state and local government and industry<br>Projects: as above plus Commonwealth government | Regional programme: state and local government and industry<br>Projects: as above |

*Figure 6.2* Key elements in the evolution of the governance and resourcing of Healthy Waterways (Source: D. Tarte, 2010)

(RWQM). Predictions from this model provided evaluation of water bodies and the basis for the SEQ Water Quality Strategy released in 1998. Model outputs also helped local councils to plan investment in upgrading wastewater treatment plants to reduce pollution.

From 1999 to 2002 (Figure 6.1) the 'SEQ Study' was broadened to incorporate rivers and inland catchments and the challenge of how to assess the water quality impacts of non-point source sediment and nutrient pollution. Again a staged approach was employed. This saw collaboration extended to include more scientists and broader public participation (Figure 6.2). Measures introduced included a cost-effective regional estuarine, marine and freshwater Ecosystem Health Monitoring Programme (EHMP), and a catchment-focused decision support tool, the Environmental Management Support System (now called 'Source Catchments'). A regional study of sediment sources was undertaken and a regional sediment transport model was developed, contributing to the calibration of 'Source Catchments' and prioritisation of further studies. Then in 2001 the 'SEQ Study' joined with the BRMG to form the Moreton Bay Waterways and Catchments Partnership, and a regional water quality management strategy was published.

The partnership was then able to enter a new phase of development. From 2002 to 2005 (Figure 6.1) activity first centred on integrating management of Moreton Bay, its estuaries and inland waterways into an Adaptive Management Framework. Decision support tools, research programmes and collaboration arrangements were integrated into an adaptive management cycle that includes data collection, analysis, planning, plan implementation, monitoring and evaluation.

From 2005, integration was deepened and collaboration further extended with the introduction of the Healthy Catchments–Healthy Waterways initiative, aimed at setting sustainable loads for pollutants entering waterways that would maintain ecosystem health. By 2005, significant progress had also been made in implementing the 2001 SEQ Regional Water Quality Management Strategy through investments in upgrading wastewater treatment plants, stormwater management programmes by local governments, riparian restoration projects and investment in the partnership's regional science, monitoring and communication work programmes. In 2005, the partnership also received funding from the Commonwealth to develop a Water Quality Improvement Plan for Moreton Bay, building on the implementation of the 2001 regional strategy. At this time the partnership was renamed the South East Queensland Healthy Waterways Partnership (SEQ HWP).

Planning intensified and the SEQ Healthy Waterways Strategy (SEQHWS) was introduced in 2007 to run until 2012. Comprised of twelve integrated action plans the strategy was developed collaboratively by partners through consultation, research and analysis. The action plans contained over 500 actions designed to improve water quality (SEQ HWP, 2007). The strategy also integrated with other relevant current plans including the SEQ Regional Plan and the SEQ Natural Resource Management Plan which set out regional natural resource targets. It was endorsed by the Commonwealth under its National Water Quality Management Strategy as the Water Quality Improvement Plan for Moreton Bay. Thus from 2009, and with this integration with higher-level plans, the emphasis shifted towards 'whole-of-water' cycle management, which includes actions to prevent water scarcity under conditions of future climate change and population growth (Figure 6.1).

Throughout this development the partnership was never a formalised legal entity although its contractual and financial responsibilities were supported by the Brisbane City Council, a key partner who also hosted the partnership's small secretariat of professional staff. During 2009–2010 new governance arrangements were agreed by partners to provide a legal structure for operation of the partnership and better linkages with regional NRM policy and governance. Thus, HW is now a not-for-profit, non-governmental, membership-based organisation that helps people work together across multiple sectors based on the belief that regional collaboration across the whole water cycle is the key to improving waterway health (Healthy Waterways, 2013). It is a collaborative network of member organisations and individuals collectively referred to as the Healthy Waterways Network who work together under the terms of the Healthy Waterways Network Rules. There are three categories of membership: investing, contributing and general. The Network is supported by the former SEQ HWP secretariat, now constituted as the not-for-profit Healthy Waterways Ltd (more commonly known as the HW Office), working under the oversight of an independent board. A Healthy Waterways Network Committee consisting of representatives from 'investing' member organisations advises this board on strategic direction and activities.

## Approach and tools

As the capacity of the partnership and an integrated approach to water management developed through the evolution describe above, it became essential for a common vision and setting of goals to be agreed. Prior to 2002 (Figure 6.1) the focus of activity had been on technical studies to characterise problems, generate data and develop decision support tools. Then through extensive consultation and as part of the development of the 2001 regional water quality strategy a common vision emerged:

> By 2026, our waterways and catchments will be healthy ecosystems supporting the livelihoods and lifestyles of people in South East Queensland, and will be managed through collaboration between community, government and industry.
>
> (SEQ HWP, 2007, p. 13)

This vision continues to guide both research and management actions, and HW's strategic approach is underpinned by two principles:

- a commitment to working in a coordinated partnership structure in which all partners can be heard, contribute to decision-making and implement agreed actions within their own spheres of responsibility; and
- formulation of management strategies on the basis of sound science, rigorous monitoring of the waterways environment and adaptive learning.

In practical terms these principles facilitate the setting of specific targets for water quality which are pursued through adaptive management and partnership working.

To support network members and achieve its vision HW delivers four key programmes. The 'Science and Innovation' programme provides independent scientific advice, develops innovative decision support tools and ensures that rigorous science underpins all activity. The 'Ecosystem Health Monitoring' programme delivers one of the most comprehensive marine, estuarine and freshwater monitoring programmes in Australia. This highlights whether the health of SEQ's waterways and Moreton Bay is improving or declining, and provides insight into the issues impacting on waterway health. The 'Water by Design' programme enables individuals and organisations to achieve sustainable urban water management through capacity-building, guidance and collaborative, science-based policy development for best-practice water-sensitive urban design. Lastly the 'Communication, Education and Motivation' programme develops and implements initiatives that engage and educate the community in the issue of waterway health, and motivates individual and collective community action (Healthy Waterways, 2013).

Over time HW has developed a sophisticated approach to identifying, measuring and managing critical pollution issues within the region's catchments. In the phase 2002–2005 (Figure 6.1) an integrated management strategy was introduced, underpinned by an Adaptive Management Framework involving five inter-linked stages (Plate 1):

- improving understanding of the environment;
- integrating science and policy planning;
- plan implementation;
- monitoring;
- evaluation.

Endorsing the notion of adaptive management (Holling, 1978), the approach is described as a process of:

> ongoing monitoring and evaluation, leading to improvement in the identification and implementation of management actions … Not only does this lead to improved understanding of ways of dealing with resource management issues, it also provides the flexibility necessary for dealing with changing socio-economic or socio-ecological relationships.
>
> (SEQ HWP, 2007, p. 61)

Another key principle incorporated into this framework is 'precaution', whereby it is recognised 'that action can seldom be postponed until we have *enough* information to fully understand the situation' (SEQ HWP, 2007, p. 60). Meanwhile, the inherently adaptive nature of the Healthy Waterway's approach means it can reflexively respond to changes in environmental and socio-economic contexts within the catchment. In order to implement this approach, a number of decision support tools have been developed within the overall planning cycle.

Several data-collection and modelling techniques have been developed and employed in gaining an understanding or characterisation of the natural environment and of the threats to waterways. For example, an erosion model (SedNet) was created to understand the spatial patterns of soil erosion processes and thus the sources of sediment entering the waterways and Moreton Bay. In the mid and upper catchments soil erosion of agricultural and grazing land results in sediments entering the river system and increasing phosphorus, nitrogen, carbon and sediment loadings, and in-channel deposition. The application of the model showed that channel erosion is a significant cause for concern in large areas of the region.

A wide range of other data are collected and analysed including physical and chemical water quality data, nutrient processing and primary productivity indicators. Wastewater nitrogen is also analysed to determine the extent of plumes in Moreton Bay, and between 2000 and 2007 a long-term research programme was undertaken in collaboration with stakeholders into the life cycle and presence of *Lyngbya* cyanobacteria blooms in the bay. Such data is used to build up an understanding of pollution causes and effects. This scientific evidence is then integrated into policy and investment planning processes to identify management actions; where possible those that are 'win–win' insofar as environmental improvements can be achieved without significant net socio-economic costs. A key decision support tool developed collaboratively with stakeholders for this purpose is 'Source Catchments'. Its main purpose is 'to evaluate the relative efficacy of various catchment management actions aimed at the improvement of water quality' (SEQ HWP, 2005, p. 190). Four sets of management options – land use change and land management practices, riparian vegetation establishment, diffuse urban stormwater management, and point source pollutant management – can be evaluated by this model, and predicted impacts aggregated across sub-catchments in the region. Resulting scenarios for catchment management have also been integrated with water quality predictions provided by the RWQM. Used in combination the models allow assessment of alternative management options and can guide stakeholders in determining sustainable pollution loadings and achievable water quality targets for the catchment.

This planning process provides a credible and sound basis for implementation of necessary strands of activity. One strand is that individual partners engage in 'on-the-ground' actions to reduce pollution inputs. For example, wastewater treatment plants upgrade their treatment to reduce the loadings of nitrogen and phosphorus in their discharges, and landholders and local government undertake riparian management utilising best practices in erosion control and farming. A second key strand consists of information campaigns aimed at publicising water quality status and management to the public and key actors. Communication of Healthy Waterway's aims through stakeholder engagement and capacity building is considered critical to the overall success of plan implementation. Various forms of outreach are utilised including websites, publications and events such as the annual River Festival (now combined with the Brisbane Festival) and International River Symposium held biennially in Brisbane. Complementary to these strands is the ongoing commitment to scientific research undertaken collaboratively with stakeholders, and monitoring of outcomes as an essential feedback and source of learning in the adaptive management framework.

Early in the development of HW stakeholders identified the need for a comprehensive monitoring programme to understand the condition of waterways, trends in their 'health' and the effectiveness of investments made in tackling pollution. Consideration was also given to how best to present scientific data from monitoring in a form understandable by all stakeholders. The EHMP developed in stages from 1997. Building collaboration between stakeholders, most notably Commonwealth and state agencies, universities, other scientists, local governments, industry and local communities came first. Data needs were scoped, and monitoring commenced for the region's estuaries and Moreton Bay, followed by the major rivers. In 2008, a rainfall event-based monitoring programme to record stream flows and pollutant loads was implemented. Combined with the freshwater, estuarine and marine components this completed a comprehensive ambient and event monitoring network for the region. This broadening of scope has required expansion of collaborative working, particularly with local governments, and a diverse range of partners now contribute to the EHMP's region-wide monitoring.

The EHMP relies on selected indicators. Five freshwater indicator groups are monitored biannually at 135 sites: fish, aquatic macro-invertebrates, ecosystem processes, nutrient cycling, and selected physical and chemical parameters. The bay and estuarine monitoring consists of 254 sites sampled every month for water quality indicators, plus annual sampling for other indicators that include wastewater nitrogen, seagrass depth range, riparian vegetation, coral cover and *Lyngbya* growth. In addition event-based monitoring records pollutant loads derived from storm events and rainfall. Analysis of this monitoring data, its evaluation and reporting closes the loop of the adaptive management cycle. It is also used to better calibrate and validate model outputs. Scientific evaluations of progress for each water body, and the reasons for improvement or deterioration, are considered with stakeholders to determine how plans require modification. Each cycle of planning should thus be based on sound evidence with regard to ecosystem health outcomes.

Widespread public reporting of monitored outcomes is an outstanding feature of the HW approach. An Annual Report Card of Aquatic Ecosystem Health (Plate 2) provides the leading means for this. First introduced in 1999, the Report Card provides an accessible visual representation and summary written explanation of the state of the region's waterways. It integrates the data from monitoring by assigning a 'score' to each sub-catchment or waterway on a scale from 'A' (near pristine) to 'F' (highly impacted). This annual 'snapshot' of waterway condition, independent of local management authority and responsibilities, is widely disseminated (Box 6.2).

---

**Box 6.2  The 'power' of the Annual Ecosystem Health Report Card**

The Report Card release is the most important annual media event for HW. It is prepared by the HW Office with quality assurance by the Scientific Expert Panel (SEP). The Report Card is released simultaneously at four well-advertised public events across the region. These are attended by ministers and local government mayors, representatives of the SEP, and community and industry partners. Because of the high level of media and public interest in the results the HW Office is asked to brief key decision-makers including ministers and mayors on the results prior to the release. This demonstrates that the public reporting of ecosystem health outcomes can become an influential part of the political process that ultimately determines prioritisation and resource allocations for environmental improvement. A further benefit for HW is that media reporting includes lead stories for TV, radio and newspapers, providing 'free' media coverage and public outreach worth an estimated AUS $400,000 each year. The Report Card is also disseminated directly by the HW Office via its website, email and public distribution of hard copies.

---

The Annual Report Card is the exemplar, but not the only example of communicating to stakeholders and involving them in evaluation and planning. Using a range of printed and online material and media presentations HW presents scientific data using techniques designed to engage non-scientific audiences. For example, visually attractive computer-generated conceptual models in graphical form are used to visually communicate water quality issues. Other techniques employed include workshops, meetings and media events. For example, workshops were held during the development of 'Source Catchments' in which stakeholders defined the characteristics required for this decision support tool, and after its development training workshops enabled them to use it.

## Getting things done: programmes and governance

### Current leading programmes and projects

Two initiatives – the Water by Design programme and the Healthy Country project – have underpinned the integrated approach for 'whole-of-water' cycle management adopted by HW since 2008. The Water by Design programme aims to build capacity in the SEQ water and urban development sectors to protect the region's waterways. Following the focus in the 2001 regional strategy on point sources, HW worked from 2005 to 2007 to scope how best to reduce urban diffuse pollution. The 2007–2012 strategy includes a Water Sensitive Urban Design (WSUD) Action Plan with the target that 'by 2026, all developed urban land in SEQ will meet consistent regional standards for WSUD'. To assess and promote the values of WSUD, HW prepared a business case summarising its economic, social and environmental benefits (Box 6.3).

---

**Box 6.3  Costs and benefits of water-sensitive urban design**

The benefits of applying WSUD to achieve stormwater management design objectives can be expected to outweigh the costs for typical low-density residential development in Queensland. The estimated average costs of applying WSUD within such residential developments equate to 0.7 per cent of a house and land package worth AUS $400,000, and could usually be passed on to the homeowner without significantly impacting the profitability of development. Estimated average annual maintenance costs are relatively insignificant at AUS $30/year, and where councils undertake the maintenance of WSUD assets in public areas these costs can be passed on to homeowners via rates.

Considering quantifiable benefits alone, the average value of total nitrogen reduction is worth more than the total life cycle cost of WSUD measures. Waterway rehabilitation costs avoided are worth around a further two-thirds of these life-cycle costs, and potential property premiums are worth around a further 90 per cent of the capital cost of WSUD. Taken together it was concluded that the potential quantifiable benefits are likely to outweigh the costs of WSUD (Water by Design, 2010).

---

The Healthy Country project is financed by the state government, contributing to the action plan in the 2007–2012 strategy that aims to reduce rural diffuse pollution by 50 per cent. Commencing in 2008 the four-year and AUS $8 million project supports communities, farmers and scientists to collectively find ways to reduce sediments and nutrients entering waterways. Action was prioritised in three catchments – the Logan and Bremer Rivers and Lockyer Creek – as these contribute the majority of sediment to Moreton Bay. There are four Healthy Country sub-projects that demonstrate coordinated responsibilities and collaboration:

- HW provides the science and planning.
- SEQ Catchments (a community-based, not-for-profit, regional NRM organisation) undertakes waterway restoration.
- Department of Economic Development and Innovation, Primary Industries Division, leads sustainable land management.
- SEQ Traditional Owners Alliance manages traditional landowner engagement.

For the first of these the Healthy Country science consortium is a partnership between the Australian Rivers Institute (Griffith University), eWater Cooperative Research Centre and HW. The consortium has prepared waterways rehabilitation plans for the three priority areas which include determination of sediment budgets, testing the budgets, and development of rehabilitation scenarios involving the activities necessary to achieve a 50 per cent reduction in sediment loads. These plans have been developed in consultation with SEQ Catchments and Healthy Country local area committees that include landowners and managers from the priority areas. Box 6.4 illustrates how the Healthy Country methodology provides advice on priority areas for erosion and sediment control interventions, and on cost-effective control measures.

---

**Box 6.4  Understanding sediment export in the Knapps Creek catchment**

- There are 38 kilometres of gulley erosion, estimated to have produced a total of 820,000 tonnes of sediment.
- Gulley and stream bank erosion are predicted to input approximately 5,950 tonnes of fine sediment to the stream network each year and hill slope erosion another 450 tonnes per year. As approximately 175 tonnes per year is deposited in the lower catchment, the net amount exported from the catchment is predicted to be of the order of 6,225 tonnes per year.
- Radionuclide concentrations in fine sediment collected from the catchment confirm the dominance of gulley and channel bank erosion.
- Results show sediment supply is dominated by sub-catchments in the middle reaches of the catchment. Ten highest yielding sub-catchments cover 10 per cent of the area but supply about 76 per cent of the sediment load.
- In summary, 90 per cent of the sediment exported from the whole catchment comes from 20 per cent of its areas located in the middle reaches.

- Management action to reduce sediment delivery needs to focus on re-vegetating and stabilising gullies in these middle reaches, and on improving and protecting riparian vegetation along channels.
- Most sediment will be delivered during large infrequent floods and management strategies should be sufficiently robust to cope with erosion and sediment loss during such events.
- The top three sediment yielding sub-catchments generate 50 per cent of the loading from 5 per cent of the catchment area, and rehabilitation of gullies and channels in these sub-catchments should be targeted as the first priority.
- Four rehabilitation methods are recommended including:
  - fencing out gullies to control stock access;
  - installation of porous weirs;
  - extensive planting of vegetation;
  - off channel water points for stock.

*Source: SEQ HWP, 2009.*

### *Working with and influencing higher-level policy*

Successful implementation of the regional strategy for water quality and action plans depends on the effectiveness of the coordination, collaboration and funding achieved by the HW Network and its partners. In turn this depends on the prevailing legal and policy environment. The HW Network must operate within complex multi-level governance arrangements, affected by decisions taken by international, national, state and local actors (Table 6.1).

At the international level several agreements are relevant to regional water quality strategy (Table 6.1). At Commonwealth level, these agreements are implemented through the Environment Protection and Biodiversity Conservation Act 1999, administered by the national government. The Commonwealth has also introduced several national coordinating policies with a bearing on catchment management in SEQ (Table 6.1). For SEQ the NRM regional boundaries are the same as HW's target area and the CfOC provides a critical funding mechanism for NRM (including water/waterways initiatives).

At the state level, environmental legislation also shapes the activities of HW. Figure 6.3 illustrates this for the key legislation and policies prevailing at the time of writing. Thus Figure 6.3 indicates that regional water quality strategy should integrate with the region's water strategy, water legislation, SEQ Regional Plan and SEQ NRM Plan, all of which are informed and shaped by state legislation. Other state legislation including measures for managing fisheries, land clearance, nature conservation, pest management and marine areas must also be considered in the HW approach. Separate state policy exists for water planning and efficiency, wetland conservation and bird protection.

*Table 6.1* Multi-level legal and policy instruments relevant to catchment management in SEQ

| Level | Legal or policy instrument | Focus |
|---|---|---|
| International | The Ramsar Convention | Protecting wetlands |
| | CAMBA | Bird protection |
| | JAMBA | Bird protection |
| | Bonn Convention | Wildlife protection |
| | UN Convention on Biological Biodiversity | Biodiversity preservation |
| Commonwealth legislation | Environment Protection and Biodiversity Conservation Act 1999 | Environmental protection, implementing international law |
| Commonwealth policy | National Action Plan for Salinity and Water Quality (NAP) | Salinity |
| | COAG Water Reform Framework and National Water Initiative | Sustainable water management |
| | National Principles for the Provision of Water for Ecosystems | Ecosystems |
| | National Water Quality Management Strategy | Water quality |
| | Natural Heritage Trust (NHT)/ Caring for Our Country (CfOC) Commonwealth Wetlands Policy 1997 | Natural Resource Management Wetlands protection |
| State legislation and policy | Environmental Protection Act 1994 | Ecologically sustainable development |
| | Water Act 2000 | Sustainable water planning |
| | Sustainable Planning Act 2009 | Land use planning |
| | Coastal Protection and Management Act 1995 | Coastal zone management and planning |
| | Fisheries Act 1994 | Sustainable fisheries |
| | Vegetation Management Act 1999 | Land clearance |
| | Nature Conservation Act 1992 | Designation of protected habitats |
| | Marine Parks Act 2004 | |
| | Land Protection (Pest and Stock Route Management) Act 2001 | Pest management |
| | Queensland Water Plan 2005–2010 | Water management |
| | Rural Water Use Efficiency Initiative | Water efficiency |
| | 1999 Strategy for the Conservation and Management of Queensland's Wetlands | Wetlands |
| | DNRW Policy for the development of ponded pastures | Habitat protection |
| | Great Artesian Basin Sustainability Initiative | Water resources |
| | Moreton Bay Shorebird Initiative | Bird protection |
| Local | Land use planning policy | Land use and management |
| | Local by-laws | Land clearance |

Source: Environmental Defenders Office (Qld) (2007).

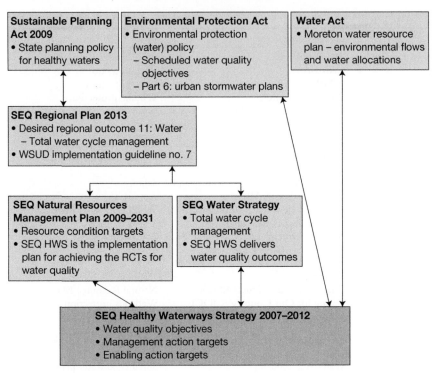

*Figure 6.3* Healthy Waterways and the prevailing policy framework (Source: D. Tarte, 2010)

Meanwhile, at the local government level, planning and local by-laws are used to protect the environment. As a consequence strategic and action plans developed by HW must negotiate and articulate strategies and actions over at least a five-year timeframe in ways that accord with and support delivery at each of these levels. As always they must also be underpinned by sound science, predictions of waterway health under proposed management scenarios, and ecosystem outcome monitoring.

The need for its work to be well-matched and well-coordinated with existing policy and legislation at all levels could be seen as constraining for a non-governmental network driven by environmental values. It certainly imposes costs in terms of the time, information, planning requirements and professional capacity required. However, it also provides opportunities. Demonstration that a network and its partnerships can be the most cost-effective delivery mechanism for a given policy objective can bring both political support and funding. It is notable that HW has achieved a track record of performance and a level of standing and acceptance such that it can be not only a trusted partner for policy implementation but also a major source of influence in shaping future Commonwealth and state policy.

## The Healthy Waterways Network

A diverse range of government and non-government partners are members of the HW Network, making it a leading example of collaborative catchment management, and one worth understanding in terms of the different responsibilities they accept, and functions or tasks they perform. At Commonwealth and state level the key role is to provide a facilitating policy environment. Government agencies are not part of HW's governance but facilitate by providing policy objectives, funding (via competitive grants) and environmental legislation and standards which legitimise HW's programmes. Technical assistance and advice is provided by CSIRO, the Commonwealth's scientific body. State government agencies similarly provide legislation, planning frameworks and funding, and technical assistance that contributes to implementation of action plans and to monitoring.

At regional level the SEQ NRM body, SEQ Catchments, was established under the NHT to develop investment strategies based on regional NRM planning, attracting and disbursing funding to local initiatives. As noted, its designated NRM region coincides with HW's target area and the two organisations work together. HW provides scientific direction and capacity-building activities and SEQ Catchments facilitates local delivery of management improvements. Seqwater, the state's statutory authority for water supply across the region, is also a key partner in contributing to implementation of relevant actions and monitoring. Policy and programme coordination is also facilitated by the Queensland Department of Environment and Heritage (previously Department of Environment and Resource Management) in partnership with the HW Office.

The key roles for local government are funding of local measures, implementation by its own staff, and contributions to monitoring. Local government responsibilities for land use planning, development guidance and approvals, health and environment, and water supply infrastructure (reticulation and treatment) are all critical areas for coordination and action.

Non-government actors are also numerous. Industry provides key partners for specific action plans and projects, community groups implement much of the work for waterway conservation and rehabilitation and contribute to monitoring, and the academic and research sector is critical in establishing the scientific consensus and credibility that underpins all actions. Local universities undertake characterisation, monitoring and reporting of environmental issues, while leading academics contribute to the guidance and evaluation provided by the SEP.

This complex network of actors is coordinated by the following governance structure. The HW board of directors has primary responsibility for governance and management, setting the strategic direction for the network and ensuring it acts in accordance with the objectives set out in its constitution. Board members are elected to serve for a four-year term by the network membership. As noted above, the HW Network Committee advises this board and seeks to ensure a balanced and integrated regional approach. The HW Office acts as the hub of the network, coordinates its programmes, supports the SEP and, where appropriate, facilitates on-the-ground implementation with local government, industry, community groups and landowners.

Effective leadership is critical in coordination and guidance of collaboration between network members. Three forms of leadership have been apparent in the development of HW. Political leadership, such as by the Lord Mayor of Brisbane, was formative in the early evolution of the SEQ HWP, and support from local and state government remains essential. Scientific leadership has been maintained through the status and authority of the SEP, and through emphasis on innovation and best practice by leading scientists in modelling, monitoring and communication technologies. Managerial leadership is provided by the network's governance structure, facilitating collaboration between members at each stage of the adaptive management framework.

Individuals who have acted as important intermediaries can also be identified throughout the network. Managers and scientific coordinators in what is now the HW Office have been central, but they also depend upon key collaborators in other organisations. As noted above, there has been explicit recognition of the need to ensure that working practices facilitate development of personal relationships and long-lasting trust between parties.

### Funding and resources

Financial resources are essential in any environmental management initiative, but the ability to draw in resources from a diverse range of sources has also helped to develop the capacity and integrated approach employed by HW. As a voluntary collaborative network it can only develop its capacity and deliver local and regional work programmes through contributions from partners and external sources.

Funding is leveraged from multiple sources and governance levels. Financial support to implement programmes has been provided by local councils and by the state and Commonwealth governments. The Queensland state government has been the most important investor in the network, providing both core funding for HW coordination and funding for implementation of the 2007–2012 strategy, which was also supported by the Commonwealth's CfOC Coastal Catchments Initiative programme. The Commonwealth has also provided grants directly to local governments, industry, SEQ Catchments and community groups for projects that implement the strategy. As HW has grown, resources have been gained from an ever-wider range of sources. These include co-investment in research and monitoring with universities, businesses and government agencies.

Network members also undertake management actions and improvement programmes at local level with resources they derive from one or a combination of the following three sources:

- as part of their core business or statutory responsibilities and thus existing operational and capital budgets (e.g. reductions in point source nutrient loads by industry and wastewater treatment plant operators, and stormwater management by local government);
- through special self-funded initiatives (e.g. waterways restoration programmes by local government);
- through voluntary community-based catchment management actions and projects.

## Measuring success: outcomes

As explained above, HW has become a collaborative environmental improvement network with multiple partners, scientific programmes and action plans that has developed over more than twenty years and measuring its effectiveness is no simple task. Natural systems take time to respond to improved management while remaining vulnerable to continued pressures and shocks, while comprehensive water quality and ecosystem health monitoring is costly. Given natural variability and other potentially causal factors such as a reduction in pressures (e.g. from closure or re-location of industry) attribution of any observed changes to actions inspired or carried out by HW may be impossible or may require a lengthy time series of data. Recognising these challenges, the definition and measurement of success has been tackled in three ways by HW.

Firstly, evidence can be analysed for single pollutants with clearly identifiable (usually point) sources. Thus, for example, an early achievement of the SEQ HWP was the success of initiatives to address pollution from wastewater. Scientific studies undertaken prior to the development of the 2001 SEQ Regional Water Quality Management Strategy identified that Moreton Bay was a nitrogen-limited embayment, and the strategy recommended improved nitrogen (5 mg/L) and phosphorus (1 mg/L) standards for discharges from wastewater treatment plants, with priority given to reduction of nitrogen. Existing discharges typically carried 30–50 mg/L of nitrogen so this required significant technological improvements. Responding to the evidence that nitrogen reduction was critical for the health of the bay, wastewater treatment managers designed plant upgrades to maximise nitrogen reduction with commensurate reductions in phosphorus. This achieved cost savings as further phosphorus reduction requires additional processes in plant design and operation. Water quality monitoring in Moreton Bay has since provided the evidence that the extent of areas impacted by wastewater discharges has been greatly reduced in size and severity. Plate 3 shows the reduced extent of areas in Moreton Bay impacted by wastewater discharges (red and pink areas) from 1998 to 2008.

The plant upgrades were also important to later supply water to a series of advanced water treatment plants built in 2008–2010 to 'drought proof' SEQ against predicted reduced rainfall with climate change. These advanced plants produce high-quality water suitable for industrial and agricultural uses and drinking water supplies. The infrastructure built provides for the addition of the treated water to water storages to increase drinking water supplies in drought conditions, an example of 'whole-of-water' cycle management.

The second and main approach adopted by HW to demonstrate outcomes has been the presentation in the annual Report Card of quantitative indicators of the quality of individual waterways and sub-catchments. An overview of results since comprehensive monitoring began suggests a mixed picture. In the 2007 strategy document (SEQ HWP, 2007, p. 27–30) both positive and negative trends are visible. While a number of estuaries and parts of Moreton Bay are in good to excellent condition (A or B grade), several estuaries and the western and southern zones of Moreton Bay either remained in poor health or deteriorated between 2000 and 2006 due to increasing population pressures along the coastline and ongoing sediment loads from the catchments. Similarly, the quality of some freshwater bodies showed improvement while slight deteriorations occurred for others. In 2009 water quality in Moreton Bay was heavily impacted by a period of high rainfall (following several years of drought) which flushed large amounts of sediment and nutrients from its catchments. Prior to this, overall water quality in the bay appeared to have stabilised. This is encouraging given that the development pressures of population growth and urbanisation have continued. Stabilisation of water quality when pressures on the environment are increasing is an achievement, but one that indicates the difficulty of the continuing challenge. Wet years in 2010–2011 and another major flood in January 2011 further confounded the picture by again mobilising large amounts of sediments and nutrients into the bay. In 2012 the bay improved in grade (C– to B–), returning closer to its long-term average grade following three years of poor grades associated with heavy rainfall. The improvement encouragingly resulted from an increase in water clarity and a decrease in nutrients and algae, suggesting improvement in inland catchments. The health of seagrass beds and corals also appeared to have returned to pre-flood conditions. This was unexpected as the 2011 flood deposited large amounts of mud on the bottom of the bay. The bay's recovery from such an extreme weather event is encouraging and highlights both the resilience of the bay and the earlier trend towards stabilisation of its water quality (Healthy Waterways, 2013). In January 2013 floods again led to increased deposition of sediment and nutrients in the bay, stimulating growth of algae, and compared to 2012 a decline in the average grade for the bay to C.

The third focus of evaluation has been to assess social impacts of the HW approach by seeking to gauge change in attitudes and behaviour, and whether broad-based societal concern and capacity for environmental improvement is really being built. Market research with the community has been undertaken to gauge awareness and attitudes towards water issues in the catchment, and the results are being used to further develop communication and outreach activities. On the whole, community perceptions of HW are positive, and its activities achieve a high level of recognition. Maintaining this community support for conserving the waterways will be a critical factor in the coming years as the population of SEQ expands further.

Another indicator of the capacity for change is the ongoing commitment of partners to the network and its programmes. Establishment of more formalised governance arrangements in 2010 indicates a long-term commitment to the vision and programmes of the network by its members, recognising the need for legitimacy and sustained resourcing in order to be able to maintain momentum and respond to emerging challenges.

## Concluding reflections

HW began as a response to water quality problems in its region and has evolved over more than twenty years in its scope, capacity, institutional structure and effectiveness. It has progressively become scientifically, spatially, collaboratively and operationally more integrated, thus developing a holistic approach to catchment management. Ultimate goals are still to be reached but HW is producing positive outcomes for protection of water quality and other natural resources by its stakeholders. The following factors and approaches have been essential for this to occur.

- Construction of a strong science-based consensus, knowledge base and capacity for targeted scientific research to address issues requiring management action.
- The broad-based engagement of stakeholders including the public.
- Strategy and action planning based on sound understanding of the waterways and developed and implemented through public consultation and involvement, incorporating the responsibilities and commitments of all levels of stakeholders.
- Continual development of trust between all HW Network members and other stakeholders.
- The articulation of a common vision underpinned by clear goals.
- Recognition of the importance of leaders and intermediaries.
- Adoption of an adaptive management framework.
- The securing of resources and a diverse funding base.

The need to establish the science and characterise the environment became evident at an early stage in HW's development. It was initially apparent that a lack of scientific understanding of the impacts of pollution provided an excuse for political inaction. Engaging partners and other stakeholders, understanding their needs and moderating their perceptions were all facilitated by robust scientific research, modelling and communication. An early initiative that has demonstrated continuing importance, value and influence was the formation of the SEP drawn from local and national research organisations and key partners. This gave the scientific programme and its tools credibility. Of equal importance was the investment made in communication tools[1] and the generation of knowledge with stakeholders. HW has invested heavily in both the natural and social science of water management (understanding, modelling, monitoring and evaluation) and the collaborative generation of knowledge with stakeholders as a basis for learning and adaptive management.

All elements of HW's approach have been inherently reliant on the approaches employed for stakeholder engagement. Development of modelling tools, strategies, action plans and monitoring programmes has all been conducted with stakeholder groups, providing a clear focus for management actions that have the ownership of governments, industry and community. When applicable, decision support tools are used with and by stakeholders to help characterise problems, identify solutions and build ownership for implementation. Catchment and receiving water quality models are used both strategically for planning, and operationally for design of infrastructure and management improvements. Community outreach, an important part of engagement, was initially under-resourced with reliance on a targeted approach, but recent years have seen more funding available for wider-reaching community initiatives.

HW has sought to build trust among collaborators as a means to reduce costs and create ownership of the process, hence helping to ensure successful outcomes. In relation to stakeholder and partner engagement individual relations have been recognised as vital. Networking and sustained personal contacts and working relationships are recognised to build both credibility and trust. Honesty, openness and being 'upfront' with people have been recognised as values important in allowing collaborative relationships to develop. At the local level the focus has been on building trust with stakeholders in the catchment, most notably farmers, industrialists and the community. Locally decision-making can depend on key individuals or elites, and activity included targeted engagement of individuals to build inclusivity, working relationships and trust.

HW staff also engaged as intermediaries at the regional, state and Commonwealth levels to develop collaboration and trust between organisations. Environmental management in Australia operates through a complex multi-level set of agencies and inter-agency competitiveness was found to present problems and limitations. Actors often worked in 'silos' of narrow responsibility and getting agencies to trust both each other and the HW approach was a challenge addressed by practising inclusivity, open and transparent dialogue and establishing collaborative working between both individuals and organisations. Consensus has been developed through inclusive committees and advisory groups.

As HW's capacity and an integrated approach to catchment management developed it became essential for a common vision and setting of goals to be agreed. Establishment of a common vision and approach has allowed the setting of specific targets for water quality and their pursuance through an agreed adaptive management approach. This had to accord with higher-level policy and planning frameworks so that local and regional activities are 'joined up' and coordinated. Targets are set within its water quality strategy by HW to deliver targets established in higher-level NRM, land use planning, water supply and environmental protection policies and legislation. This provides legitimacy as well as coherence, and can lead to funding. Leaders who provide political, scientific and managerial leadership have been essential, particularly in the early stages of initiatives and programmes. Intermediaries or 'network champions' have also been very significant assets.

HW has sought to integrate best-available scientific knowledge with stakeholder engagement in an adaptive and collaborative approach. The adaptive management framework adopted has been based on comprehensive characterisation of pollution sources and the receiving environment, collaborative planning and plan implementation, ongoing monitoring of impacts, science-based evaluation and public reporting of results. Communication and outreach in relation to outcomes aims not only to disseminate results but also to increase stakeholder understanding of the effectiveness of management options. This is essential in generating the trust and commitment required of, and between, partners for sustained delivery of measures that can be difficult and costly. The aim is also to elicit and integrate local knowledge with scientific data, so as to enhance location-specific understanding of pollution problems and the applicability of solutions. Setting all of this within the adaptive management framework also provides for both the network and partners to learn through critical self-reflection.

Monitoring of water quality is central to the necessary scientific research and to adaptive management, and again various partners contribute to catchment-wide ecosystem health and event monitoring. Broad-based trust in these processes enables scientific evidence and risk assessments to become the basis for policy, planning and management processes. The annual Report Card (and associated achievement awards) showcase both successes and areas for improvement, and have real influence on the attitudes and behaviour of policymakers. Drawing in resources from multiple sources and levels of government helps to build integration and coordination. Next year's budget is never certain, but diverse funding can build confidence and a strong degree of financial sustainability when results can be demonstrated and effectively reported.

Ultimately the success of the HW Network will only be demonstrated by achievement of the vision of healthy waterways and ecosystems supporting the livelihoods and lifestyles of the region managed through multi-level and cross-sectoral collaboration. In the interim, success is being demonstrated in terms of water quality and the capacity of the people, processes and institutional arrangements being built. Physical monitoring shows that environmental pollution impacts in Moreton Bay have been stabilised over the past decade despite the developmental pressures of one of Australia's fastest-growing regions. Social research indicates that HW is contributing significantly to the sense of well-being and confidence in the future of the region's residents, while educating them in their roles, responsibilities and expectations for their environment.

## Note

1　Area maps and other detailed catchment information and communications tools can be viewed from the Healthy Waterways website (http://www.healthywaterways.org/Home.aspx).

## References

Commonwealth of Australia (1999) *Managing Natural Resources in Rural Australia for a Sustainable Future: A Discussion Paper for Developing a National Policy*, Department of Agriculture, Fisheries and Forestry, Canberra.

Davie, P., Stock, E. and Low Choy, D. (eds) (1990) *The Brisbane River: A Source-Book for the Future*, Australian Littoral Society/Queensland Museum, Morooka, QLD.

Department of Environment and Resource Management (2011) *Queensland Regional Natural Resource Management Framework*, Department of Environment and Resource Management, Queensland Government, Brisbane.

Environmental Defenders Office (Qld) Inc. (2007) *Overview and Analysis of the Legislative and Institutional Framework relating to the SEQ Healthy Waterways Strategy Part 1*, Environmental Defenders Office, Brisbane.

Farrelly, M. and Conacher, A. (2007) 'Integrated, regional, natural resource and environmental planning and the Natural Heritage Trust Phase 2: A case study of the Northern Agricultural Catchments Council, Western Australia', *Australian Geographer*, vol. 38, no. 3, pp. 309–333.

Great Barrier Reef Marine Park Authority (2009) *Great Barrier Reef Outlook Report 2009*, Great Barrier Reef Marine Park Authority, Townsville.

Gregory, H. (1996) *The Brisbane River Story: Meanders Through Time*, Australian Marine Conservation Society, Brisbane.

Healthy Waterways (2013) *Working Together to Improve Waterway Health*, www.healthy waterways.org/Home.aspx, accessed 20 August 2013.

Holling, C. S. (ed.) (1978) *Adaptive Environmental Assessment and Management*, John Wiley & Sons, London.

McDonald, G. and Roberts, B. (2006) 'SMART water quality targets for Great Barrier Reef Catchments', *Australasian Journal of Environmental Management*, vol. 13, no. 2, pp. 95–107.

NSW Government (2013) *For Catchment Management Authorities*, www.environment.nsw. gov.au/4cmas/, accessed 19 August 2013.

Park, G. and Alexander, J. (2005) 'Integrate or perish – lessons in integrated NRM in North Central Victoria', *Australasian Journal of Environmental Management*, vol. 12, supplement 1, pp. 47–56.

SEQ HWP (2005) *Healthy Waterways, Healthy Catchments*, South East Queensland Healthy Waterways Partnership, Brisbane.

SEQ HWP (2007) *South East Queensland Healthy Waterways Strategy 2007–2012: Strategy Overview*, South East Queensland Healthy Waterways Partnership, Brisbane.

SEQ HWP (2009) *Rehabilitation Priorities Knapp Creek – Final Report*, South East Queensland Healthy Waterways Partnership, Brisbane.

Water by Design (2010) *A Business Case for Best Practice Urban Stormwater Management* (Version 1.1), South East Queensland Healthy Waterways Partnership, Brisbane.

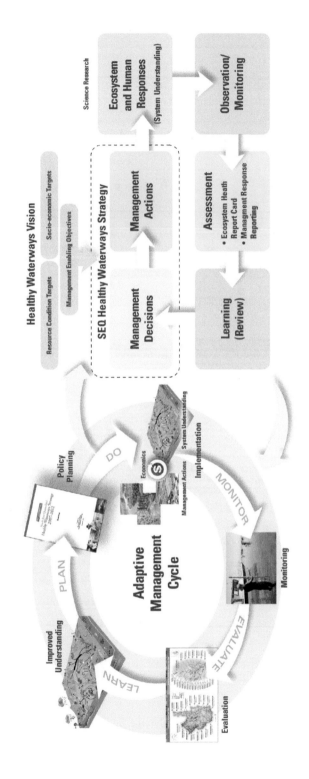

*Plate 1* The Healthy Waterways Adaptive Management Framework (Source: South East Queensland Healthy Waterways Partnership, Brisbane, Qld, Australia)

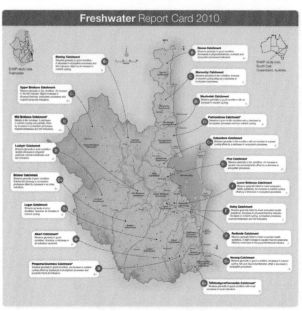

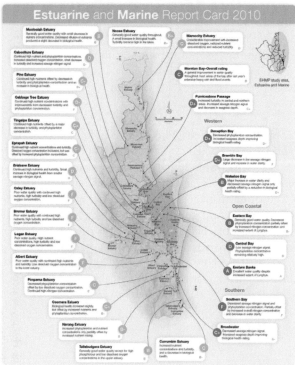

*Plate 2* The 2010 Ecosystem Health Report Card (Source: South East Queensland Healthy Waterways Partnership, Brisbane, Qld, Australia)

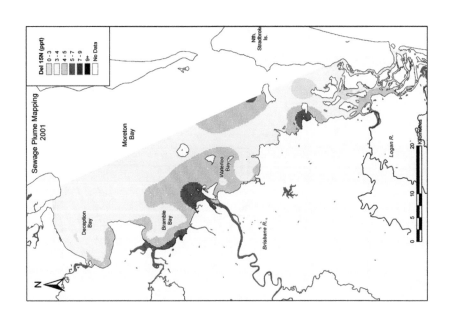

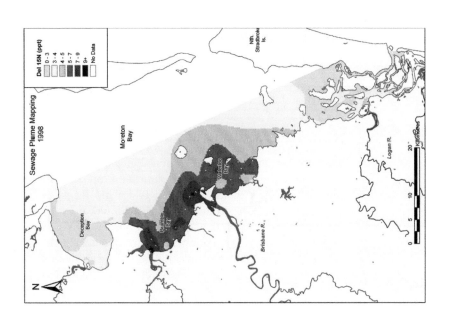

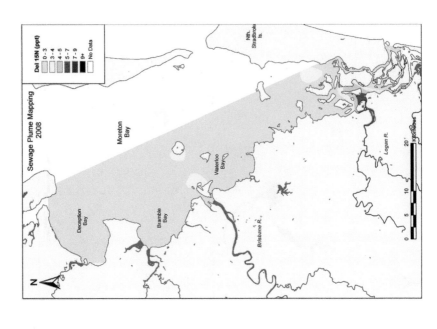

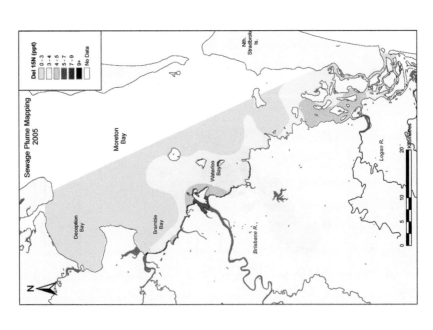

*Plate 3* Wastewater plume mapping in Moreton Bay, 1998–2008 (Source: South East Queensland Healthy Waterways Partnership, Brisbane, Qld, Australia)

# 7 Groundwater protection programmes in Denmark, Germany and the Netherlands

*Laurence Smith, David Benson and Kevin Hiscock*

## Introduction: the challenges of groundwater quality

The other case studies in this book focus on surface water catchments, but we should not neglect the importance of groundwater, nor the lessons that can be learnt from initiatives for its management and protection. Groundwater quality is important for the safety of drinking water and for the health of aquatic ecosystems, but pollution prevention, water quality restoration and monitoring are more difficult to achieve for groundwater than surface water. Groundwater is 'hidden' and inaccessible and it requires expertise and resources to locate and quantify the scale and severity of pollution. Restoration of the quality of groundwater, once degraded, can be difficult and may require a long timescale.

In Europe, agricultural intensification over more than 60 years has caused increased leaching of nutrients and pesticides to groundwater and surface water, and in areas with vulnerable (permeable) soils and aquifers pollutant concentrations may often approach or exceed limits set for these contaminants. While agricultural sources are of prime concern they are not the only diffuse or dispersed point sources. Residential septic systems, garden usage of chemicals, hydrocarbons from transport networks and leaching from old landfills, mines, quarries and other industrial sites can also be significant.

Among all pollutants affecting groundwater, nitrate presents the greatest challenge because of its abundant sources and chemical characteristics. The natural level of nitrate in groundwater is usually below 10 milligrams (mg) of nitrate ($NO_3$) per litre and higher levels are caused by human activity. Mineral fertilizers account for almost 50 per cent of nitrogen inputs to agricultural soils in European farming systems, animal manure for 40 per cent, and biological fixation and atmospheric deposition the remainder (Aue and Klassen, 2005). Nitrate is soluble, very mobile and persistent under aerobic conditions. Under the European Union (EU) Drinking Water Directive (Council of the European Communities, 1998) a standard of 50 mg $NO_3$ per litre (mg/l) has been set for drinking water based on the World Health Organization's guidelines for drinking water and the opinion of the European Commission's Scientific Advisory Committee (European Commission, 2013a). The EU Nitrates

Directive (Council of the European Communities, 1991b) is an integral part of the EU Water Framework Directive (Council of the European Communities, 2000) and has the objective of protecting water quality across Europe by preventing nitrate pollution of ground and surface waters by agricultural sources. Surface water and groundwater are identified under this directive as polluted, or at risk of pollution, if the concentration of nitrate similarly exceeds 50 mg/l (European Commission, 2013b). Many regions in Europe are finding this to be a stringent target, and thus recognize diffuse and dispersed point sources of nitrate pollution as among their most important water management challenges.

Approximately 75 per cent of the population of the EU depends on groundwater for water supply (Aue and Klassen, 2005). It is also an important resource for industry and for irrigation. It is, however, increasingly recognized that groundwater is not just a water supply reservoir but a critical ecosystem within the hydrological cycle. Wetlands and surface flows depend on groundwater particularly when rainfall and runoff is lowest. Many streams and rivers receive a significant proportion of their flow from groundwater and thus its quality directly affects their water quality and aquatic ecosystems.

When groundwater is polluted a short-term 'technical fix' for drinking water can be to deepen wells to abstract from older, uncontaminated layers. This increases costs, may drawdown a water table reducing flows to surface waters, and can only work where geological and hydro-geological conditions are favourable (Aue and Klassen, 2005). Common alternatives are 'end-of-pipe' water treatment and blending of supplies of differing quality. Such measures do not address the causes of pollution or protect aquatic ecosystems, and are also likely to face increasing costs (including energy costs and greenhouse gas emissions) but they do offer some certainty for the safety of drinking water supply.

The 'catchment management' alternative is to protect water quality at source through land use and management options that protect groundwater. A key mechanism for this approach in some EU member states has been to establish cooperative agreements between farmers and water suppliers. These are invariably voluntary agreements, often with the direct or indirect support of local authorities (Heinz, 2008). Key features of such mechanisms are that they: are written agreements based on the mutual interests of the parties involved; involve water suppliers acting as initiators and, usually, the main source of finance; are targeted within groundwater catchments or protection zones; and rely heavily on self-regulation for enforcement (Heinz, 2008). Examples exist in at least nine EU countries but such agreements have been most prevalent in Denmark, Germany, the Netherlands and France (Brouwer et al., 2003), where implementation is often integrated with spatial and land use planning at a local or regional scale. In this chapter, three case studies illustrate how such a catchment management approach has been developed and applied.

## Context and drivers

A number of directives have provided regulatory drivers for the protection of groundwater in EU member states. Since December 2000 all of these directives have become progressively subsumed under the aims and required measures of the Water Framework Directive (WFD). The Nitrates Directive (91/67/EEC) aims to reduce water pollution by nitrate from agricultural sources and to prevent further such contamination. It is intended to protect groundwater and to prevent eutrophication of inland and near coastal waters. It requires action to be taken where surface water or groundwater nitrate concentrations exceed 50 mg/l or could exceed this limit if no action is taken, and where a land area has been identified as contributing to pollution and thus designated as a Nitrate Vulnerable Zone (NVZ). The Nitrates Directive remains in force as an instrument for achievement of the aims of the WFD, and requires member states to establish voluntary codes of practice for farmers to protect waters from nitrate pollution. The Groundwater Directive (2006/118/EC) (Council of the European Union, 2006) aims to protect groundwater from toxic, persistent and bio-accumulative pollutants. Once such contamination has occurred remediation or treatment costs can be high. A risk assessment and prior authorization are therefore required before disposal of specified substances onto land, according to the legislation of each member state. The Groundwater Directive sets out specific criteria for the identification of significant and sustained upward trends in pollutant concentrations, and for the definition of starting points for when action must be taken to reverse these trends. In this respect, significance is defined both on the basis of time series and environmental significance. Time series are periods of time during which a trend is detected through regular monitoring. Environmental significance describes the point at which the concentration of a pollutant starts to threaten to worsen the quality of groundwater. This point is set at 75 per cent of the quality standard or the threshold value defined by member states. The Plant Protection Products Directive (91/414/EEC) (Council of the European Communities, 1991a) provides for authorization of relevant pesticide products within the EU and defines criteria and the process for safety assessment of active ingredients. The Drinking Water Directive (98/83/EC) sets standards for public health and drinking water supply post-treatment.

The WFD (2000/60/EC) establishes a legal and administrative framework for protection of all water bodies (inland surface, groundwater and near coastal) based on common objectives, principles and overall approach for all member states. Core objectives are to prevent deterioration of aquatic ecosystems and restore all waters, including groundwater, to 'good status' as defined by ecological, biological, physical and chemical parameters. As noted, the directives above specify common measures that contribute to achieving this 'good status'. The WFD requires a cyclical process of River Basin Management Planning, with improvement in water quality to be achieved through the implementation of River Basin Management Plans (RBMPs) for each of the designated River Basin Districts. RBMPs contain programmes of measures (POMs) designed to achieve the objectives and are implemented and reviewed on a six-year cycle. It is intended that these plans should provide a means for public and stakeholder participation in the implementation of the directive. This ethos is intended to identify and exploit 'combinations of measures' that can address river basin problems at different spatial scales, and gain the synergies from combining regulatory, voluntary and incentive-based approaches. It provides opportunity, legitimization and standing for non-government initiatives, such as by water suppliers and farmers, that contribute to WFD objectives.

The Common Agricultural Policy (CAP) is also relevant to the regulatory context for groundwater protection. Under the CAP farmers that receive subsidy must also comply with a range of statutory management requirements. Such 'cross-compliance' includes land management requirements defined under the Nitrates Directive for NVZs. The reduction in subsidy for non-compliance provides an incentive for good practice, but monitoring and enforcement of 'cross-compliance' is costly and time-consuming. The implementation and effectiveness of this policy varies across member states, although rigorous evaluations of non-compliance are generally lacking (Alliance Environnement, 2007). In addition to the principles of 'cross-compliance' it is accepted under the CAP that provision of environmental services by farmers beyond a common baseline level merits incentives. 'Agri-environment' schemes have implemented this principle since the late 1980s in support of specific farming practices. Farmers receive payments that compensate for the additional costs and loss of income incurred. Although there are similarities with cooperative agreements between water suppliers and farmers, 'agri-environment' schemes are usually nationwide in implementation, funded under the CAP and to date have had limited application for surface water or groundwater protection measures.

In each case study in this chapter the agencies responsible for water supply and environmental protection seek to achieve the aims of the Nitrates and Water Framework Directives and to provide safe drinking water supplies. They face common challenges of nutrient and chemical pollution of groundwater and recognize that solutions require land use and management measures well-matched to groundwater catchment conditions and in accordance with local and regional spatial plans and national and EU policies.

### Denmark

In Denmark drinking water is obtained almost entirely from groundwater. The water supply industry is highly decentralized and made up of approximately 2,700 private but not-for-profit operators, 200 public waterworks and 90,000 private wells that each supply from one to nine households (Aue and Klassen, 2005). The main threats to be managed are pesticides and nitrates in vulnerable areas.

Factors supporting groundwater protection at source are national policy that favours this compared to 'end-of-pipe' treatment and the pride that localities take in their ability to deliver untreated potable water. The Planning Act of 1992 also requires economic and spatial plans at regional, municipal and local level to be unified, providing a framework for the planning and coordination of groundwater protection. For example, unified regional and municipal plans for the County of North Jutland and the City of Aalborg summarize and prioritize activity and investment by sector. The regional plan sets out guidance for land use in the open countryside that includes the principles that: unpolluted groundwater should be the basis for drinking water supply; preventative action and protection of important and vulnerable drinking water areas has high priority; and a decentralized water supply system should be maintained. Land areas are categorized for three levels of priority for protection for drinking water abstraction, and areas are designated for afforestation. In the municipal and matched local plans future land use is similarly specified.

Further to this the Groundwater Protection Act in 1998 provided for formulation of Groundwater Protection Plans (GPP) for areas declared as catchments for the abstraction of drinking water. The legislation is founded on Danish Legislation on Environmental Objectives (*miljømålsloven*), Danish Legislation on Water Supply (*vandforsyningsloven*), the Danish Planning Act (*planloven*) and subsequently on Danish adoption of the EU WFD (Aue and Klassen, 2005). A GPP should be based on a detailed survey of the hydro-geological conditions, a specification of land use and assessment of all known existing and potential sources of pollution. Surveys are undertaken to reveal where groundwater is most vulnerable and efforts need to be concentrated. The GPP should describe what actions need to be taken, designate the agency or agencies responsible and provide a timescale for implementation. Zoning will then take place for well fields and groundwater catchment areas, with designation of target areas for action to control nitrate, pesticides and other contaminants. Since implementation of the EU WFD the GPP and associated measures have become part of the River Basin Management Plans (Aue and Klassen, 2005).

A leading local example of groundwater protection is provided by the Drastrup Project of the City of Aalborg. This project preceded the introduction of GPPs but from its start the Municipality of Aalborg applied a then unique unified planning approach that used planning legislation to plan and change land use with the aim of protecting groundwater. Sixty per cent of the city of Aalborg's 120,000 inhabitants are supplied with water by the Aalborg municipality through a commercial, but non-profit, public utility company that competes with private water suppliers. The source groundwater is abstracted from three catchment areas to the south of the city and supplied untreated. In these areas neither the subsoil nor underlying chalk layer provides significant natural filtration to limit percolation of nitrate and pesticides to the groundwater. In the 1980s it was identified that water quality was threatened. Monitoring recent groundwater recharge showed a mean nitrate content in excess of 120 mg/l and traces of pesticides (Aue and Klassen, 2005). It was recognized that without effective measures the groundwater would become permanently polluted. It was also recognized that only 6 per cent of the municipality was forested compared to a national average of 12 per cent, and that an increase in forested area could deliver benefits for recreation and biodiversity. In 1986 the Drastrup area, ten kilometres from the centre of Aalborg, was selected as a pilot area for groundwater protection.[1] It was known that the selected area had to remain a continuing source of water supply for the city despite its vulnerability to nitrate and other pollution. Dual objectives were to develop new methods to sustain abstraction of safe untreated drinking water and to expand recreational areas in a location accessible to city residents.

## Germany

In Germany, over 70 per cent of drinking water is obtained from groundwater, and this rises to over 85 per cent in the state of Lower Saxony, located in the north-eastern part of the country. It is recognized nationally that groundwater needs to be protected as drinking water and to prevent eutrophication and preserve aquatic habitats. As in Denmark, water supply is decentralized within over 7,000 water supply companies, mostly owned by communities or municipalities (Aue and Klassen, 2005). The most significant pollution threat to groundwater is from intensive livestock farming and diffuse leaching of nutrients from fertilizer and animal manure.

Regulation in Germany generally requires water quality standards that equal or exceed those required by EU directives. Details of the regulatory and implementation approach tend to vary between federal states (Länder) reflecting their constitutional autonomy for implementation of national framework laws, but should be in accordance with national guidelines published and updated since 1953 by the German Technical and Scientific Association for Gas and Water (DVGW). Under national law, provision also exists for 'Water Protection Decrees' (WPD) to be made for water protection areas.

Normally a water protection area covers the total surface water and groundwater catchment area, and is divided into three zones. In Zone I the objective is protection against direct contamination and other impacts within a radius of ten metres around an abstraction well. Zone II extends from Zone I for the distance over which infiltrating rainwater will take 50 days to reach the abstraction well. For this zone the objective is protection against pathogenic microorganisms and impacts from human facilities and activities including land use. Zone III extends from Zone II to the boundary of the groundwater catchment area, with the objective of protection against persistent impacts of chemical and radioactive substances (Aue and Klassen, 2005).

For each zone, WPDs can define rules for land use and management, and impose penalties for non-compliance. When defining such restrictions the authority responsible must consider the existing intensity of agricultural, urban or industrial land use, all sources of pollutants including atmospheric deposition, and local climatic, soil and geo-hydrological conditions. It is expected that the standards and limits defined should be at least superior to the 'cross-compliance' standards of good agricultural practice under the EU CAP and Nitrates Directive, effectively representing the maximum tolerated emission for farming activities consistent with safe drinking water provision. Such stringency can cause conflicts of interest, for example, through imposing spatial limits on industry or constraints on farming. WPDs can also provide an edict for 'cooperation' between farmers and water suppliers. In Lower Saxony, for example, measures were urgently needed to protect drinking water supplies from nitrate pollution and about 14 per cent of the land was declared a water protection area (Aue and Klassen, 2005). A 'cooperation decree' then made provision for farmers to be offered free advice on land use and farming practices and compensation payments for protection of groundwater defined through negotiation and funded by a supplementary water abstraction charge paid by water consumers.

The Groundwater Protection Programme of the Water Board of Oldenburg and East Frisia (OOWV) in Lower Saxony has taken advantage of these institutional arrangements (Aue and Klassen, 2005). OOWV is one of the largest water suppliers in Lower Saxony, supplying drinking water from groundwater to approximately one million water consumers. Its southern catchment areas are subject to high leaching rates and intensive agriculture as farmers compensate for low fertility sandy soils with high stocking rates and intensive use of organic manures and inorganic fertilizers. This situation led to high and increasing nitrate concentrations in production wells in the 1980s. Wells were deepened as an urgent measure to reduce the nitrate levels, but to sustain a long-term ability to supply high-quality drinking water required OOWV to develop a comprehensive groundwater protection programme in partnership with the then district administration of Weser-Ems.

A local example is provided by OOWV's Thülsfelde Waterworks and Water Protection Area, located 30 kilometres south-east of the city of Oldenburg. This source produces 11 million cubic metres of drinking water per year from 40 abstraction wells. Since 1978 groundwater has been abstracted from a depth of 30 to 160 metres and the catchment has a surface area of about 7,360 square kilometres. Annual rainfall of 819 mm, a water balance of 263 mm, and homogeneous sandy soils are conditions that support water abstraction but are also vulnerable to nitrate leaching. In the 1980s it was observed that intensive agriculture and animal husbandry systems were resulting in high nitrate and pesticide concentrations in the production wells. To ensure drinking water standards OOWV had to close affected wells and seek a long-term solution (Aue and Klassen, 2005).

### The Netherlands

In the northern province of Drenthe in the Netherlands groundwater supplies 95 per cent of drinking water, but high levels of abstraction can have negative environmental impacts (Aue and Klassen, 2005). There are extensive nature reserves, and in particular wetlands with rare vegetation types dependent on high groundwater levels and constant seepage from underlying aquifers. Projects to manage groundwater in this region have therefore aimed to counteract any lowering of water tables and protect against nitrate and other pollutants. As in Denmark and Germany, developing solutions in the Netherlands relies on the ability to implement changes in land use and management.

Policy objectives and implementation in the Netherlands must also similarly accord with the relevant EU directives. National policy additionally decrees that to prevent flooding of low-lying urban areas District Water Boards should select floodplain areas for controlled flooding, temporary storage and slow release of flood water during extreme rainfall events. Responsibilities for implementation coalesce at the level of provincial government which has responsibility for policy and planning for the environment, including the quality and quantity of surface water and groundwater. It is recognized that achievement of re-naturalization of floodplains and wetlands, reduction in downstream flood risk, and protection of surface water and groundwater quality will require development and imple- mentation of an integrated spatial plan.

For example, in Drenthe Province relevant provisions have been integrated in one document, the Provincial Integrated Spatial Plan or 'POP'. Local municipalities then prepare their own spatial plans to be 'nested' within this provincial planning framework. For Drenthe the 'POP' has a particular focus on maintaining a safe and sustainable drinking water supply, combined with aims of re-establishing wetland habitats in areas where they once existed and reducing flood risk. This planning process provides the framework that allows initiation of integrated water resource management projects that include groundwater protection among their objectives.

A local example is provided by the Hunze Valley, a lowland river flowing through the north-eastern region of Drenthe Province with a very suitable aquifer for drinking water abstraction. Originally meandering over a shallow gradient, the river had been canalized since the 1950s to more rapidly drain farmland, and its water level in the lower reaches became higher than the land surface. In some places old meanders are retained as valuable remnants of the river's natural morphology, but elsewhere they have been incorporated into farmland. Many low-lying farming areas remain prone to flooding.

## Getting started

### The Drastrup Project, City of Aalborg, Denmark

After selection as a pilot groundwater protection area in 1986, the Drastrup groundwater protection project was launched in 1992. Covering an area of approximately 900 hectares it aimed to sustain abstraction of drinking water and expand forest area for city residents. The start of this initiative can be attributed to the vision of the Aalborg City Council and its planners and water managers. Its origins lay in the development of a unified plan for future land use in the area. This plan was the first of its kind in Denmark (WaterCost, 2008), and was holistic in scope despite being driven by the need to sustain safe drinking water abstraction.

In 1995, the land use plan for the area was officially adopted and a working group began formulation of an EU 'LIFE' project application. 'LIFE' is the EU's financial instrument supporting environmental and nature conservation projects, and Aalborg's planners recognized that the Drastrup scheme could provide a precedent in a European context. From 1996 the City Council started to provide funding for the purchase of land and this accelerated when support was gained from EU 'LIFE' in 1997. In September 2001, the 'LIFE' project came to an end with the opening of a new recreational forest within the established groundwater protection catchment.

### The Groundwater Protection Programme of the Water Board of Oldenburg and East Frisia (OOWV)

As noted above, occurrence of high nitrate and pesticide concentrations in abstraction wells in several regions of Germany led to legislation in the federal states affected that provided for groundwater protection measures financed by a water abstraction levy. Actual provisions vary by state in terms of the amount of the levy and the degree to which it is hypothecated, or 'ring fenced', for sustainable land use and water protection. The 'Water Law of Lower Saxony', which also regulates implementation of the EU WFD, stipulates that water suppliers are responsible for the monitoring of surface water and groundwater in water supply catchments to ensure sustainability of drinking water quality and quantity.

By 1992, the State of Lower Saxony had recognized that nitrate concentrations in groundwater had risen and it consequently introduced a water abstraction levy (the 'water penny'). Domestic water consumers paid a levy of five cents per cubic metre of water, with lower rates applying to large industrial and agricultural users (for example, the nuclear power industry paid a levy of one cent per cubic metre of cooling water). This revenue is transferred from the water company to the state government, and 40 per cent is hypothecated for measures for sustainable land use and water protection (Aue and Klassen, 2005).

It was understood by policymakers that a solution to the nitrate problem could only be developed by working with all land users. Thus Cooperation Committees were formed for each waterworks and their water supply area and comprised representatives from farming, horticulture, forestry, water companies and other relevant organizations, chaired by the local district council. The role of these committees was to collaboratively develop solutions and the Groundwater Protection Programme of the Water Board of Oldenburg and East Frisia (OOWV) is a leading example.

Measures to reduce pollution had in fact been first implemented in the late 1980s but from 1992 the 'water penny' could be used to finance further groundwater protection. The finance enabled the development of a combined programme led by OOWV and the district administration in cooperation with local partners and stakeholders.

### The Hunze Valley, Drenthe Province, Netherlands

The results of geo-hydrological surveys showed that the aquifer system in this area offered good hydraulic characteristics and high quality groundwater (Aue and Klassen, 2005). It was thus identified as very suitable for drinking water production, and when the local drinking water company applied for a licence to produce drinking water from a new groundwater production field in this area, their proposal was judged to be compatible with the provincial and municipal integrated spatial plans. However, the planned production field initially provoked strong opposition from local farmers who feared the groundwater protection rules that would accompany a new well field. In particular they feared constraints imposed on current farmed area, future expansion of farmed area that might otherwise offer advantages of economies of consolidation and scale, and potential restrictions on their farming practices. It was thus evident that the water company and the local administration needed to engage with all stakeholders and to develop a process of cooperation that could research and consider options acceptable (and if possible beneficial) for all affected parties. This cooperation has since formed the mechanism that facilitates a series of water management projects and improvements.

## Approach and tools

### The Drastrup Project, City of Aalborg, Denmark

Central to the approach of Aalborg City Council was the development of a unified plan for future land use in the Drastrup area that was holistic in its objectives and implementation. Thus the plan included: a strategy for protecting groundwater; sustainable land use through forestry; permanent grassland; 'environmentally-friendly' farming; countryside areas for recreation; a ban on new sources of pollution; and elimination, where possible, of all causes of water pollution (Aue and Klassen, 2005). The municipal plan that initiated the project provided for a doubling of the forest area, an increase in biodiversity and better public access to natural areas. To secure and enhance the characteristics of the landscape, forest has been located on the moraine hills of the area while river valleys have been kept more open. Footpaths and other recreation and picnic areas have been introduced.

Land use change and the removal of polluting activities from designated areas was achieved through a unified approach to spatial planning, stakeholder engagement and partnership working with relevant organizations and local communities. Both landscape values and sector interests were analysed to develop the details of the plan, and to evaluate, coordinate and prioritize the varied interests that existed in this near-urban but still rural groundwater catchment. Such competing interests included farming, forestry, urban development, recreation, raw materials (sand and gravel), transport routes, sewer systems, waste disposal and natural habitat provision for biodiversity.

Land use change consisted of conversion from intensive agriculture to more extensive farming and forested recreational areas through voluntary and compensated agreements for change in farming practices (an example of cooperative agreements), voluntary land sales, or land swaps for areas outside protected zones. Voluntary agreements for farming practices converted areas under intensive cultivation and livestock rearing to permanent grassland with lower stocking densities or organic farming. Gravel pits and waste disposal sites were closed and their landscapes restored to forest and grassland through voluntary purchase agreements. Afforestation of such sites and former farmland with native broadleaf species has been the most commonly used approach. This strategy achieves a permanent change in land use and thereby permanent protection of the aquifer. It also had the advantage of quite quickly providing a visible landscape change that reinforced public engagement, particularly when these areas became accessible for recreation.

Another innovative aspect of the Drastrup Project was that two small towns within the groundwater protection area became the focus for information campaigns and provision of advice by the City Council. This approach called on households and small businesses to refrain from use of pesticides and other chemicals in gardens, yards and driveways. The City Council also led by example by stopping pesticide use in public spaces such as parks, squares, churchyards and playing fields. Community engagement groups set up to inform the spatial planning during the project period have been sustained as follow-up groups, enabling the City of Aalborg to continue to work collaboratively with the area's landowners, businesses and citizens.

Scientific research and evidence have underpinned decision-making at each stage in the Drastrup process. It was needed, in particular, to provide the basis for the voluntary agreements with farmers. For example, new knowledge was needed to determine the exact grazing pressure that would still ensure both clean groundwater and a high degree of biodiversity on permanent grassland in this area. Research showed that if application of fertilizer and sprays is managed, and grazing (or hay or silage removal) is balanced at the right level to remove nutrients and achieve biodiversity conservation, conversion to grassland could be a cost-effective and low-risk groundwater protection measure (WaterCost, 2008).

Development of a shared knowledge base was supported by collaborative engagement with farmers as partners, and the building of trust. Without this latter feature, it was found difficult to persuade farmers and livestock owners to accept constraints in grassland management on fertilizer and pesticide use, and on the choice of cultivation methods for re-seeding and establishment. Emphasis on long-term planning and management was also essential. Good soil structure and a diverse flora and fauna on meadows and common land can take years to build up but little time to degrade. However, it was found that groundwater protection can be obtained rapidly from sustainably managed grassland as long as residual stocks of nutrients and pesticides in the soil are not too high, and a good management regime is maintained (Aue and Klassen, 2005).

Research has also underpinned afforestation work. Partnership with the Danish Forest and Landscape Research Institute has developed methods of forest establishment that aim for the lowest possible leaching of nutrients to groundwater. Conventional planting methods were found to be problematic as soil disturbance can conversely cause mineralization, liberating nitrate and other nutrient salts to soil water with a high probability of subsequent leaching to groundwater. The greater the soil disturbance incurred, for example from deep ploughing, the greater the risk of leaching of nutrients from the soil, although alternative cultivation methods are more labour-intensive and costly. Similarly, use of manual weed control can reduce pesticide use during establishment. The project evaluated and accepted that costs in terms of reduced sapling survival, growth rate and ground cover, were outweighed by the benefits of enhanced water quality protection (WaterCost, 2008).

Forest management methods similarly required careful evaluation. The natural accumulation of nitrogen in a forest released when windfall or clear-cutting occurs is a potential risk to groundwater quality. Mitigation strategies include selective harvesting and replacement to achieve a stand of mixed age, allowing natural regeneration, and whole tree harvesting to remove all plant material (and thus the nitrogen it contains). However, even a forest laid out for natural succession can become a source of nutrients over time. In the worst case this occurs through storm damage, disease or other causes of poor growth and regeneration. Sustaining groundwater protection by afforestation thus requires sound empirical research and expertise to support evaluation of multiple criteria and complex choices of planting and management methods.

Finally, continuous monitoring of groundwater quality has been essential for two main reasons. First, it enables evaluation of the effectiveness of the land use and management changes made. Second, it provides the basis for information campaigns that sustain partner engagement and public participation and support (Aue and Klassen, 2005).

## The Groundwater Protection Programme of the Water Board of Oldenburg and East Frisia (OOWV), Germany

A founding assumption for this programme was that a solution to nitrate pollution could only be developed by working in cooperation with all land users. A shared programme of endeavour was needed to develop appropriate pollution control concepts and methods and to put these into practice. Thus Cooperation Committees were formed for each of OOWV's well fields and waterworks. These brought stakeholders from agriculture, horticulture and forestry together in direct dialogue with the water supplier and local authorities.

From the early 1990s the groundwater protection programme that developed was then based on three integrated 'pillars': cooperative agreements with farmers; purchase of land and its afforestation; and the promotion of organic farming. Public information campaigns were important in supporting this programme, and raising public awareness of environmental issues, particularly those related to the water cycle, has been of central importance. For example, teaching trails and guided tours that inform about groundwater protection, water abstraction and organic agriculture have been used with schoolchildren and adults. Scientific research and groundwater monitoring provided the knowledge base to underpin all activities. Finance has been provided by the 'water penny' supplemented by contributions from environmental foundations. Additionally, the state government makes compensation payments for lost farm production in water protection areas where agricultural cultivation is restricted under a Water Protection Decree.

### The Hunze Valley, Drenthe Province, Netherlands

In Drenthe Province, the Provincial Integrated Spatial Plan assigns drinking water production fields and their groundwater protection zones, as well as areas reserved for future drinking water production. Such reserved areas contain groundwater of good quality and sustainable abstraction can take place without a negative impact on nature reserves. Typically, they are located within the lower reaches of the catchments of small rivers, where the effects of abstraction can be compensated by the abundant surface water, and where habitats usually depend more on surface water than groundwater and thus suffer little from groundwater abstraction. A contrast is offered by habitats at the foot of hills dominated by wetland vegetation dependent on a high water table and well-buffered mineralized groundwater quality. It is thus recognized that groundwater production should not be sited on hill tops or near the foot of hills, where abstraction will effect groundwater availability and quality in the root zone (Aue and Klassen, 2005). Land use within the catchment area of a potential groundwater production field is a second important selection criterion. Ideally, the land should be a natural reserve rather than (in order of decreasing risk) an industrial, urban or agricultural area.

To achieve change in land use, and in particular to change from agriculture to other functions such as natural habitat creation and water storage, requires cooperation between stakeholders including landowners, the drinking water supplier, the District Water Board, the local municipality, the provincial government and environmental organizations. This partnership approach has sought to develop and implement integrated spatial planning in pursuit of solutions that deliver multiple benefits.

The integrated planning framework developed has now facilitated an ongoing series of projects and management interventions. These seek to maintain safe and sustainable drinking water supply combined with re-establishment of wetland habitats in areas where they once existed, reduction of downstream flood risk and productive agriculture. Trade-offs are inevitable, and for example, several drinking water production well fields have needed to be relocated within the past 25 years of integrated planning and implementation to avoid negative impact on protected nature reserves or because of vulnerability to one or more sources of groundwater pollution.

To reduce flood risk national policy identifies a sequence of necessary actions; that is, to prevent rapid discharge of water, store water and then allow controlled discharge from an area. In the Hunze Valley this policy is implemented by broadening the riverbanks and by retaining or restoring meanders. The resulting wide floodplain can provide temporary water storage and contribute to reduced flood risk downstream. Additional benefits derived include restoration of the original wetlands, improvements in water quality in those wetlands because of less need to introduce water to them in dry periods, and development of attractive landscapes for recreation and as habitat. The meandered valley of the Hunze should ultimately serve as an ecological 'corridor', providing a natural link between habitats.

In implementing land use change the water supplier has generally sought to buy the wettest lands that are least agriculturally productive. Land outside the area of their interest has also been purchased and used in land exchanges with farmers in the target areas. Landowners who continue to farm in the areas affected have recognized that the restoration of the meanders and wetlands attracts tourists. New income-generating opportunities include farm campsites and rental of canoes and bicycles. Local shops and restaurants have also experienced increased demand from the increasing number of visitors, leading to diversification and/or expansion of their business, and corresponding benefits for the local community.

Research and a strong scientific evidence base have underpinned the spatial planning and the design of specific interventions. For example, a complicating factor was the presence of an important drinking water production well field – 'De Groeve' – within the river floodplain. Research and monitoring were required to assess the potential to sustain shallow groundwater abstraction while restoring the wetland, the capacity for wetland vegetation to absorb nutrients, and the overall effect of natural flooding on surface water and groundwater quality. The resulting management interventions developed have been able to deliver an improved and lower-cost resource for drinking water abstraction, water conservation and improved water quality.

## Getting things done

### The Drastrup Project, City of Aalborg, Denmark

As noted above the principle of voluntary change was central in the Drastrup Project. Change in land use has been achieved entirely through voluntary management agreements or land swapping or voluntary sales. From 1997 to 2001 the City Council implemented a programme to redistribute the land and change its use. Farmers who wished to continue conventional farming were offered land outside the project area, and the municipality offered to buy any land that farmers wished to sell. Voluntary cooperative agreements for management practices were concluded with those farmers who continued to farm in the area. The City Council also reached voluntary agreements with the owners of two gravel pits to stop the excavation of raw materials and to carry out the post-extraction restoration of the gravel pits in a sustainable manner.

The voluntary cooperative agreements for less intensive farming were implemented through the legal instrument of 'cultivation declarations'. These are guidelines registered for a land area, usually by its owner in return for a capital payment, which establish permanent changes to the way the land is managed and continue to prevail if ownership of the land changes (Section 26a of the Environmental Protection Act; Aue and Klassen, 2005). The compensation payment, implementation period and environmental outcomes of a 'cultivation declaration' depend on the restrictions imposed. Because a capital incentive payment is made to establish the declaration they can appear costly for a publicly funded project, particularly in the near term. It is thus important that the declaration provides for a permanent solution to the environmental problem being addressed. It will also be expected that the landowner in receipt of the payment will already be compliant with existing regulations and minimum standards of good farming practice. Declarations should specify conditions which are enforceable and penalties for non-compliance, and should make provision for inspections or other appropriate monitoring (Aue and Klassen, 2005).

Ongoing scientific research supports the technical basis for 'cultivation declarations' and management strategies for the whole area. When farmland is converted to permanent grassland actual outcomes were found to depend on a number of key parameters including field topography, soil water balance and atmospheric deposition of nitrogen. Variables that can be managed in response include soil texture, grazing pressure, type of grazing animals and use of fertilizers and pesticides. The project has demonstrated that given the right conditions and management options even intensively farmed land can be restored to support a more balanced ecosystem and increased biodiversity within a few years (Aue and Klassen, 2005). These improvements in turn create a demand for public access which can be facilitated as part of cooperative agreements and through provision of pathways.

National policy has also facilitated afforestation in the Drastrup Project. The national afforestation programme has the goal to double the forest area within a 100-year period (Madsen, 2001). The Aalborg City Council chose to conform to this objective and to work to double the forest area within the municipality which had a naturally low density of woodland. Another supporting aspect of the socio-political and policy environment has been a historic tradition to protect and preserve woodlands to secure timber for building ships and for firewood. Because of this context all publicly financed or co-financed afforestation is placed into 'forest reserves' with a legal 'preservation duty'. This measure provides a level of legal protection that makes it virtually impossible to convert back to arable land (Aue and Klassen, 2005). Finally, there has been growing public interest in the use of woodlands for recreation and Aalborg, in common with many European cities, suffers from a shortage of such resources in its vicinity.

Protection of groundwater was thus achieved in the Drastrup Project through three approaches: afforestation through buying (or swapping) land; conversion to permanent grassland through buying (or swapping) land; and conversion to permanent grassland or other low intensity farming systems through 'cultivation declarations' agreed with farmers. The first and second offer both greater permanence and stronger protection, that is, given the same preconditions they will be less sensitive to external influence and variation, for example, from atmospheric deposition or flooding. In contrast, use of 'cultivation declarations' can aim only to meet prescribed water quality standards and the ownership of land remains with the farmer. Although continuing farming is less intensive, a professional farmer will still seek to optimize production and income within the constraints of the prevailing regulations and terms of the agreement. The level of protection achieved has a greater degree of uncertainty and a greater requirement for regular monitoring, as well as less permanence (WaterCost, 2008).

Afforestation of purchased land was also found to be the most cost-effective method to reduce nitrate and pesticide pollution, whereas the cost-effectiveness of conversion to permanent grassland was similar for both land purchase and cultivation declarations (WaterCost, 2008). Afforestation can also be expected to generate the most additional benefits in terms of recreation, but there will usually be a limit to the area of afforestation acceptable in any given location once local preferences and trade-offs with productive farming are considered.

Funding for implementation of the Drastrup plan was derived from multiple sources, although most important were Aalborg City Council and an EU 'LIFE' project. Clear principles informed an apportionment of costs, acceptable to all parties. The costs of realizing benefits classifiable as 'public goods' and of benefit to the wider public (for example, recreation, landscape and biodiversity conservation) were financed from taxation, while the costs of securing clean drinking water were financed from payments by water consumers. Overall financing of the plan was also supplemented from the EU 'LIFE' project.

The investments made in the Drastrup Project were substantial but Aalborg City Council has found that social cost benefit analysis demonstrates a positive return on its investment. Afforestation in urban and nearby areas that previously had a low forest density raises residential property prices in the area by an amount in aggregate that exceeds the costs of buying and afforesting land. Revenues from property taxes similarly increase. The city has been highly satisfied that afforestation delivers multiple benefits, and that the right location and afforestation methods will readily justify the land use change in terms of the environmental, economic and social returns to the investment.

The Drastrup Project was the first of its kind in Denmark, and it was exceptional in that the Municipal Planning Act was used to change and plan land use for protection of groundwater. This project therefore provides an exemplar of a long-term, holistic and participatory spatial planning approach to groundwater protection for multiple benefits. However, somewhat ironically, options for groundwater protection in Denmark have arguably become both more formalized and more limited since the 1998 Groundwater Protection Act (GPA). Since the Act, revenue from water consumers can only be used for the least extensive land management measures necessary to meet water quality targets. In most situations this means employing no more than 'cultivation declarations' and excludes the purchase of land for multiple purposes – recreation, nature and groundwater protection – as was a key feature of the Drastrup Project.

The 1998 GPA requires the development of Groundwater Protection Plans (GPP). These describe the actions to be taken to protect groundwater, including the party responsible for carrying out the initiatives, and when they are to be implemented. In addition, they must include a programme for monitoring groundwater quality, and guidelines for permits, licences and other decisions that can be taken by the authority responsible (a county or municipality). Before a GPP can be adopted it must be presented to a coordination forum consisting of representatives from the water supplier(s), local authorities and the agricultural sector. The plan is also submitted for public consultation for a period of eight weeks. In its entirety, preparation (including any necessary surveys) and adoption of a GPP requires at least three to four years but should achieve a comprehensive and holistic plan addressing all sources of pollution. Implementation of the plan is then an ongoing and adaptive process based on continuous monitoring and an evaluation once in each four-year election cycle for local authorities.

Once a GPP has been adopted all affected landowners must be informed of its implications for them. A GPP enables cultivation restrictions to be imposed on farmland to reduce pollution by nitrate and pesticides through 'voluntary' agreements based on full economic compensation for affected land users. If voluntary agreements cannot be achieved the local authority can compel change in land use, again with full compensation. All other initiatives must be carried out on a voluntary basis or under the aegis of existing legislation.

The implementation costs of a GPP depend on the initiatives in the plan, and although they can be financed from general taxation they are most commonly financed by a levy on water consumers, hence the principle that expenditure should only be made for groundwater protection and not for other benefits such as recreation or nature conservation. This has led to the tendency for 'cultivation declarations' to become the sole policy instrument for groundwater protection. Such 'cultivation declarations' are negotiated with farmers and can meet groundwater protection targets by reducing pollution, whereas in the Drastrup Project a broader range of measures including land purchase for afforestation or permanent grassland and community engagement were more cost-effective, eliminated some sources of pollution and provided other benefits (WaterCost, 2008).

As noted earlier, the water supply industry in Denmark is highly decentralized with a large number of private but not-for-profit waterworks (water suppliers), which are often operated on a voluntary basis by communities. The adoption of a GPP can also have negative effects insofar as that it can be difficult for small water suppliers to implement the actions required. Faced with the challenges of pollution and GPP requirements water suppliers have needed more professional staff. Some have closed but instances of cooperation between small private and public water suppliers have increased; usually led by the larger public water supplier and its professional staff. In the municipality of Aalborg, for example, such collaborative arrangements include a shared pipe network for mutual security of supply carrying 98 per cent of abstracted water, a shared monitoring programme, knowledge-based, shared public information campaigns, and further afforestation efforts. Such cooperation between private and public waterworks is now regarded as key to the survival of small private, community and non-profit waterworks in Denmark.

### The Groundwater Protection Programme of the Water Board of Oldenburg and East Frisia (OOWV), Germany

Since 1992 the revenue derived from the 'water penny' has been used in Lower Saxony to develop cooperation between land users and water suppliers under the Water Protection Decree and edict of cooperation. This form of collaboration has involved provision of additional advice to farmers in water catchment areas, and financing of voluntary agreements between farmers and the local administration for implementation of specific agricultural measures and special water protection projects. The water supplier, OOWV, authorized the Chamber of Agriculture to provide advice to the farmers in their water protection areas and organized pilot projects to inform the farmers about alternative less intensive and polluting forms of agriculture. Via the water supplier and the state government the water consumer and taxpayers fund the provision of advice to farmers and the adoption of the voluntary farming measures.

The voluntary cooperative agreements take the form of contracts between land managers, OOWV and the state of Lower Saxony. Under the terms of the contracts the enterprises (principally farms, but also forestry and horticultural operations) are obliged to observe restrictions and conditions on land use over and above baseline good environmental practice. For example, farmers may be required to plant maize in closer rows (45cm) and apply less fertilizer, and receive compensation for the economic loss then incurred (Aue and Klassen, 2005). In addition to compensation for income foregone, an additional incentive payment can be included where the measure is known to be effective but not in widespread use.

Such agreements are the most important component of the approach to groundwater protection in Lower Saxony. Measures can be quickly and efficiently implemented by the farmers in contrast to the design and monitoring of a regulatory approach such as prohibitions and requirements imposed by an order for a water protection area. There is a much greater acceptance by farmers of cooperative agreements because they actively participate in and influence the development and selection of the measures to be employed, because there is no coercion and because compensation is agreed. The agreements themselves are developed through collaborative working by the Chamber of Agriculture, water suppliers and local government. Working groups meet annually to update the agreements in line with the results of water quality monitoring and scientific research. A 'catalogue of measures' developed as a reference is used to streamline the negotiation process with individual farmers and to help coordinate a fair and consistent application of agreement conditions and associated payments. The agreements also have the flexibility to be adapted locally to prevailing conditions, thus achieving both better effectiveness and acceptance in specific catchment areas or water protection zones (WaterCost, 2008).

Preparatory work and advice to farmers is carried out by the Chamber of Agriculture and water protection advisors through trials and demonstrations, farm scale models and financial illustrations. This work is conducted both with groups of farmers and on an individual basis. The water company usually employs eight advisors for each water protection area. Final authority for the specification of an agreement rests with the water supplier, although all possible attempts are made to base this on consensus and mutual agreement. The 'catalogue of measures' is circulated to all land managers in the water catchment areas at the beginning of the year, with application deadlines for specific agreements. Water protection advisors canvass farmers to make contracts and assist them in carrying out the measures. Compliance with the conditions of an agreement is monitored by spotchecks. Farmers are also required to keep records but this has not always worked well (Aue and Klassen, 2005). If contraventions are found a discussion is first held with the farmer, followed by suspension of compensation payments if necessary. Repeated contraventions can lead to exclusion from the programme. Other sanctions are not provided for except in the case of EU co-financed measures. If compliance is satisfactory payments to the farmers are made by the water supplier. Such monitoring and enforcement of agreements is not easy and is potentially costly. Establishment of mutual trust is seen as critical but requires a long time perspective and effective processes of farmer and public engagement and awareness raising (Aue and Klassen, 2005).

A second 'pillar' of the programme has been the purchase of land and its afforestation. An early example of this, and the first large-scale attempt at drinking water protection through silviculture in Lower Saxony, was an afforestation project in the Northwest German Forest District of Ahlhorn, realized by a partnership between OOWV and the state forest agency. This project aimed to exclude further application of liquid manure and pesticides to the soil for a target of 1,800 hectares of afforested land (Aue and Klassen, 2005). To optimize groundwater production (by avoiding the greater and year-round evapotranspiration of conifer trees), and to reduce acid deposition (caused by atmospheric deposition of nitrogen), the forest agency committed to plant broadleaf trees. The most sensitive recharge areas in the vicinity of wells or infiltration areas were targeted for new forest, while farmers in adjacent areas were offered programmes to support adoption of farming methods which could safeguard groundwater quality. Initially, afforestation was opposed by farmer organizations and a dialogue supported by expert advice was initiated to address this.

From 1989 to 2008 afforestation of nearly 1,200 hectares of water protection area was achieved (Aue and Klassen, 2005). This land was acquired by OOWV at prevailing market prices through voluntary sales by farmers, and then assigned to the state forest agency which financed the planting and management of the trees. The possibility for further afforestation was not excluded but attention and priority then turned to the management of the existing woodlands. These remained a source of risk because of the sandy soils on which they were typically located and the high prevailing rates of atmospheric deposition of nitrogen (30–35 kg/ha/year). It was recognized that sustainable management would continue to require cooperation between local authorities, water suppliers, the state forest agency and private landowners, as well as the continuing development of management measures for groundwater protection.

As the third 'pillar' of the water protection programme organic farming has been established by scientific investigation as a recommended measure to protect groundwater against nitrogen and pesticides. Its impact derives from lower nutrient inputs, improved soil quality and a crop rotation that has a smaller percentage of maize and a higher percentage of summer grains after a winter catch crop (Aue and Klassen, 2005). Organic farming had been increasing from a low level in Germany because of support available from EU and federal programmes and increasing national demand for organic food. However, for economic and historical reasons organic farms tended to be concentrated in the eastern and southern states of Germany and not in regions with specialized and intensive production systems. The standards of the organic producer organizations in Germany are in several respects stricter than EU regulation for organic production. For example, they prescribe the conversion of the whole farm, the import of slurry from conventional farms is not allowed, and only 1.4 livestock units are permissible per hectare. Monitoring is conducted by private inspection bodies approved by state authorities. In Lower Saxony, payments financed by the 'water penny' were combined with those available from the EU for the promotion of organic farming, and farmers were provided with free advice from a dedicated consultancy.

An example of a groundwater protection project that integrates the 'pillars' of the approach described above is provided by the Thülsfelde water catchment. Here OOWV was able to buy 500 hectares of land for the purpose of converting it to more extensive land use. A significant proportion was handed over to the state forest agency of Lower Saxony for the purpose of afforestation, 250 hectares was rented to organic farms and some was used for low-intensity orchards. In 2005 the total water protection area consisted of 7,357 hectares; 55 per cent under agricultural use, 39 per cent under forest and 6 per cent as restored wetland (WaterCost, 2008). The water abstraction charge introduced in 1992 and the edict of cooperation provided the basis for an additional advice service for the farmers in the area provided by the Chamber of Agriculture, and for cooperative agreements that offered compensation payments for production foregone. Sixty per cent of farmers in the water protection area signed at least one agreement. OOWV also bought land for the purpose of changing its use from intensive agriculture to more extensive and particularly organic farming. Tenant farmers now farm organically on the land which was bought for this purpose, while OOWV retained ownership of one organic farm on the most vulnerable soils as a demonstration farm (*Biohof Bakenhus* close to the Thülsfelde protection area).

In addition to the use of voluntary instruments for groundwater protection the OOWV applied for the status of a water protection area in several water catchments. For example, in 2000 the water protection area of Thülsfelde was established and farmers were obliged by the concomitant regulations to meet specific restrictions concerning slurry application, nutrient balancing and storage of fodder and fertilizer; albeit with compensation payments for this.

Catchment advisors have played an essential role in providing advice to the farming community on groundwater protection measures. Their knowledge of the agriculture of the area enables recommended measures to be tailored to the local situation at a farm or even plot scale. In many catchments the advisor has served not only as the main planner, but also as a broker mediating between the different interests of water supply, profitable farming, biodiversity conservation and land uses. Experience showed that the success of a catchment advisor was dependent on gaining the confidence of all stakeholders (Aue and Klassen, 2005).

The functions of the catchment advisor included offering a free advisory service on groundwater protection. A focus particularly valued by farmers was advice on improvement of fertilizer management. Advisors recruit as many farmers as possible to take part in cooperative agreements, and support the realization of them in practice. Advice is provided through individual meetings, farmer group sessions and written circulars. Achievement of efficiency and cost-effectiveness is a major goal for the advisors, and thus they are involved in monitoring so that data from soil samples, field trials, nitrogen balancing and other methods can be used in the essential function of demonstrating success to the farmers, water supply companies and the general public. Farmers and other stakeholders appreciate being shown things in action before they accept and adopt new ideas, and thus demonstration farm sessions and open days have been found to attract large audiences (Aue and Klassen, 2005).

An ongoing programme of scientific investigation was instituted to inform and support these initiatives, and specifically to develop and improve the water protection measures applied. A knowledge base has accumulated from a range of studies that include: denitrification processes in the aquifer, methods of afforestation, grassland management, organic farming and cooperative agreements. This scientific programme required effective and close collaboration between scientific institutes and universities, the Chamber of Agriculture, farmers, farmer associations and the relevant administrative bodies. The results of these scientific investigations were discussed with local stakeholders and the efficacy and efficiency of measures, plus further research needs, were used as the basis for negotiations on cooperative agreements. Publishing the results of scientific investigations via the internet has made it possible to inform interested professionals and gain peer review, as well as to inform the wider public. Ongoing problems that have been identified as requiring discussion and solution by the programme partners with the support of the scientific institutes include: improving the cost-effectiveness of the measures adopted to protect groundwater (especially against a background of reform to government structures in Lower Saxony and possible constraints on financial resources), management options for mature forest stands, achievement of higher rates of farmer participation in the schemes, scientific knowledge of denitrification processes in the subsoil, and the growing threat from incentives for farmers to re-intensify arising from improved commodity prices and national policies that subsidize bio-energy production from field crops and manures.

As noted above in relation to the role of advisors, monitoring has been an essential element of the groundwater protection programme. The budget for the groundwater protection programme is continually under review, and consequently effective prioritization of measures and control of costs is paramount. In order to monitor and evaluate groundwater protection activities a wide-ranging and purposeful set of instruments and methods was established, again through consultation with stakeholders and in partnership with scientific institutes. Each monitoring parameter and method was considered in relation to its effectiveness for the examined medium, the spatial and temporal reference and resolution, and the costs (for which there was a wide range). Inevitably the core of the monitoring required is to take groundwater samples, but maintenance and use of groundwater monitoring points is relatively expensive. Fortunately, all water suppliers usually maintain a network of shallow groundwater monitoring points within water protection areas. This network reduced capital installation costs, but data collection and analysis costs are still incurred for all sampling above that which is routine for the water supplier. In addition to water samples, the potential risk of nitrogen leaching is assessed through soil samples taken in the autumn (unsaturated zone, mineral-N), although costs for this are fairly low. Where field drains exist it is possible to assess field-specific emissions of pollutants such as nitrogen, but the diversity of site conditions requires a well-elaborated and site-specific set of methods in order to gain sound and reliable data (Aue and Klassen, 2005).

## The Hunze Valley, Drenthe Province, Netherlands

Over time the Province of Drenthe has developed an integrated approach to spatial planning based on the application of clear criteria for location of drinking water abstraction and wetland restoration. The Hunze Valley provides a good illustration of how this process can work. The river's lower reaches had been canalized in the past to facilitate rapid flood discharge and the dominant land use was agriculture. Rural–urban migration had been reducing the population as younger people moved for study and employment, and consequently local shops and other services had been in decline. Policymakers at provincial and municipality level wished to reverse these trends.

To achieve both re-naturalization and reduction in downstream flood risk the integrated spatial plan for the lower reaches of the Hunze provides for wetland habitat to return and for a re-meandering of the canalized river. Geo-hydrological surveys also showed that the aquifer system and groundwater quality in this area can remain suitable for drinking water production, and be enhanced. The District Water Board readily grasped the opportunity to invest in restoration of the floodplain to contribute to its aim of downstream flood risk reduction. However, plans to expand drinking water production were initially a source of contention and provoked opposition from farmers concerned that this would impact upon their livelihoods by limiting areas that could be farmed, thus limiting potential benefits of farm expansion and economies of scale, or by restricting farming practice. High-quality technical assessment and analysis, openness in sharing information and careful spatial targeting of measures were essential in overcoming these fears. The water supplier was then able to implement its plans through land purchases (of the lowest-lying and wettest areas) and land exchanges.

The water supplier also worked with farmers still farming land in vulnerable parts of protection zones surrounding the groundwater production field. The aim was to reduce the use of pesticides and loss of nitrate to groundwater. Farmers were individually coached on best farming practices and measures, and given the small numbers selected this training could be tailored for each farm. Although this approach was relatively costly it yielded good results and a high level of commitment among the farmers involved. The approach included creation of knowledge exchange groups for agriculture in which farmers could participate without cost. Topics considered by these groups included: prevention of crop diseases (healthier crops more optimally absorb nutrients and can therefore reduce leaching of nitrate to groundwater) and prevention of nematodes to enable a green manure crop to be grown in winter (reducing leaching of nitrate in this season). On-farm demonstrations by their peers helped farmers to learn the benefits of advanced and more environmentally friendly farming practices including alternative crop rotations, crop varieties and manuring practices. Other knowledge exchange groups were established to focus on the means to meet standards for nitrate and phosphate losses from livestock farming. Advice to farmers also extended to the income-generating opportunities created by increased numbers of recreational visitors and tourists (Aue and Klassen, 2005).

## Outcomes

### *The Drastrup Project, City of Aalborg, Denmark*

The Drastrup Project is notable for its multiple achievements and beneficial outcomes, demonstrating how objectives for groundwater protection, biodiversity conservation and recreation can coincide. When it commenced, land use in the area consisted of conventional farming, operational gravel pits and two small towns. About 450 hectares of farmland have now been converted to forest or permanent grassland and along with closed and restored gravel pits have become attractive recreational areas with picnic sites and 55 kilometres of walking trails (Aue and Klassen, 2005). Such trails also connect the area to the city of Aalborg and neighbouring towns. New methods for afforestation and sustainable farming that are well-adapted to the area and achieve mitigation of nitrate pollution of groundwater have been developed, and pesticide use has ceased on all forest and permanent grassland. Residents in the two local towns have also declared their gardens to be pesticide-free. Monitoring of wells has shown improvements in nitrate and pesticide concentrations. For example, the nitrate content in the soil water has decreased to less than 10 mg/l. The costs of the project have been significant but the investment made by the City Council has been recovered from the increased rateable value of properties within and near the project area.

Despite its patient and collaborative approach, its duration and its emphasis on voluntary agreements the Drastrup Project has incurred obstacles. Compared to the original plan almost 300 hectares of farmland continues to be intensively farmed and one gravel pit is still in operation. This is because it has not been possible to conclude voluntary agreements with the landowners concerned, and the City Council has so far declined to use more stringent measures to achieve the desired change.

### *The Groundwater Protection Programme of the Water Board of Oldenburg and East Frisia (OOWV), Germany*

The groundwater protection programmes of OOWV in Lower Saxony demonstrate considerable achievements. As explained above farmers can enter into a cooperative agreement for groundwater protection with the water company for land situated within the catchment areas of wells used for drinking water. Revenue from the 'water penny' is used to compensate loss of farm income resulting from higher costs or reduced production. Specially trained advisors engaged through partnership working with the Chamber of Agriculture have encouraged as many farmers as possible to agree to these contracts and have provided free advice on groundwater protection and land use.

This cooperative and voluntary approach achieved a steady increase in participation and commitment by farmers, such that by 2008 there were 115 cooperation programmes between farmers, local authorities and water suppliers throughout Lower Saxony, covering almost all groundwater protection areas (WaterCost, 2008). This represented a total cultivated area of about 300,000 hectares and involvement by more than 6,000 farmers. Evaluation of the early years of this activity also informed development of the approach that has become the basis for groundwater protection in all water catchment areas in both the state of Lower Saxony and nationwide.

Because of the long timescales associated with groundwater recharge, the concentration of nitrate in groundwater decreases only after a delay following implementation of any protection measures. Measurable reductions in nitrate have, however, been recorded in some catchment areas. An example of this successful outcome is provided by the water protection area Thülsfelde. Here nitrate concentrations in the lower groundwater, measured in observation wells at depths of from 5 to 30 metres below the surface, show the success of the combination of measures applied, as there has been a marked decrease in nutrient and pesticide concentrations. This can be attributed to the commitment made over in excess of ten years to build cooperation, trust and patience between all participants. The continuation and further technical development of protection measures is important for success but it is the cooperative committees and effective partnership working between the agencies concerned that is the most important element.

The costs of groundwater protection have been significant but it should be noted that the state government retained 60 per cent of the revenue from the 'water penny', while the remaining 40 per cent financed the programme. This burden fell on water consumers but the cost per household or business has been relatively modest. In 2008, expenditure for groundwater protection was equivalent to approximately 10 Euro cents per cubic metre of water supplied, compared to prevailing treatment costs for the removal of nitrate of 50 Euro cents per cubic metre (WaterCost, 2008).

### The Hunze Valley, Drenthe Province, Netherlands

In the Hunze Valley integrated spatial planning and partnership working is delivering multiple benefits encompassing safe drinking water production, flood risk reduction, habitat restoration, increased tourism and farm diversification. Initial opposition to the expansion of drinking water production was overcome when the water supplier was able to demonstrate that it would buy only the areas of least agricultural value or could offer land swaps for productive areas beyond the boundaries of the scheme. The District Water Board exploited the flood management benefits from restoration of the river floodplain, and local communities including farmers benefited from increased recreational and tourist activity. This helped to reverse the trend of closures of local shops and other businesses, and cut migration. It has been recognized that the inclusive and participatory planning approach backed up by sound technical evidence, and the emphasis given to tourism developments, were essential to gain the support and ownership of the scheme by local stakeholders. This overcame initial scepticism and opposition, and created both willingness and commitment for future protection of the riverine environment.

## Concluding reflections

### *The Drastrup Project, City of Aalborg, Denmark*

The Drastrup Project provides an exemplar of a long-term, holistic and participatory spatial planning approach to groundwater protection. Although protection of drinking water supply was the initial driver, it was soon recognized that an integrated approach could generate multiple benefits. Despite potential conflicts of interest between different stakeholders, the project's implementation was facilitated by a number of factors. Firstly, the urgent need to protect groundwater for drinking water supply is widely accepted in Danish society, and the national political context and policy framework favours long-term protection of water at source rather than investment in 'end-of-pipe' treatment. Secondly, a facilitating legislative, planning and political environment coincided at municipal and regional levels. Thirdly, Aalborg municipality had, and further developed, the technical capacity for implementation of a scheme of this nature through its own staff and partnerships with other organizations. Fourthly, benefits from biodiversity conservation, recreation and increased property values in addition to groundwater protection garnered broad-based support from a wide range of stakeholders in the area.

Locally, the fact that implementation of the integrated plan for the Drastrup area stemmed from a democratic decision taken by the City Council is given great importance. This factor and the processes of consultation and public participation that took place ensured that all stakeholders could have their viewpoints heard before final decisions were taken. Contact and cooperation with all stakeholders in the area via civic associations and other organizations, and their involvement in the process of integrated spatial planning were essential to project success. This applies not only for design and implementation, but also to 'anchor' and sustain desired outcomes in perpetuity (Aue and Klassen, 2005). The plan itself was based on a rigorous scientific landscape characterization and analysis made understandable to lay citizens. Public information and awareness-raising campaigns were sustained and effective. Thus on two counts, scientific credibility and support from local civil society, the City Council gained a sound basis for the decision to proceed.

This decision-making was not, however, a one-off process. Implementation of the project has involved a continuous process of follow-up planning and analysis, accompanied by continued dissemination of information and dialogue with people affected. As a result many examples of cooperation between the City Council and local citizens can be observed: for example, public participation in monitoring wildlife and flora, and the self-declaration of two small towns as pesticide-free (following the example set for public spaces by the City Council and advice it provided).

Overall it can be concluded that the Drastrup Project was initiated and carried through because it was possible to resolve and bring together competing interests to support action at a time when the legislative framework and political commitment within the municipality were also enabling and supportive. The project was a unique approach, based on voluntary agreements and land swapping, which may no longer be possible in Denmark on the same scale given current interpretation of the Groundwater Protection Act (WaterCost, 2008). This is despite the fact that there is widespread agreement that the project has achieved multiple benefits beyond clean drinking water. It was an approach that was very applicable for a relatively small catchment area in proximity to an urban centre, driven by an urgent need to protect drinking water and with strong demand for recreational access to forest and permanent grassland. It may be less applicable on a larger scale and in more remote rural areas where broad-based environmental conservation and water quality protection are the primary drivers.

### The Groundwater Protection Programme of the Water Board of Oldenburg and East Frisia (OOWV), Germany

The main elements of this programme are regulation, inter-agency cooperation, voluntary cooperative agreements with farmers, promotion of organic farming and purchase of land for afforestation; all supported by scientific research and public outreach. In Germany the Water Protection Decrees that facilitate such a programme are considered an effective means to protect groundwater, but only if the restrictions placed on farming are reasonable and appropriate for the location, compensation payments are available, a trained advisory service exists to support the farmers and the authority responsible has the legitimacy and legal standing to be able to resolve conflicts if necessary.

The process requires an active and effective intermediary to promote and coordinate action. A team leader is needed to liaise between the state governments, institutions and farmers. This role is usually played by the water supplier, driven by their need for improved water quality protection. The cooperation of farmers is critical but can be difficult and take time to achieve. Essential to this cooperation are quality data to demonstrate the scale and severity of the problem, finance to pay for improved farm practice and broad-based public support. There needs to be a clear geographical focus on which to base assessments and cooperation, which is best provided by the water protection zones designated around wells. The experience of the OOMV in Lower Saxony well-illustrates these points.

Some uncertainty regarding the sustainability of groundwater protection programmes has, however, arisen in recent years. From 1992 to 2005 the 'water penny' facilitated concentrated and effective cooperation between

OOWV, the Chamber of Agriculture and the relevant district administrations. Public concern for environmental protection and support for water protection at source rather than 'end-of-pipe' treatment was strong, and an additional levy of 5 Euro cents per cubic metre of water consumed had wide acceptance. However, the district administrations were annulled as part of fiscal reform in 2004–2005 and since then local administrative responsibilities with respect to groundwater protection have been less clearly defined. From 2008 water suppliers have been given responsibility for managing expenditure from the 'water penny' and have agreed to continue water protection schemes. Success in improving groundwater quality over 20 years has also come under threat of reversal as improved commodity prices and national policies that subsidize bio-energy production from field crops and manures have provided incentives for farmers to intensify, reducing the attractiveness of cooperative agreements and raising the costs of compensation. The financial resources of municipalities and states for groundwater protection have also come under much greater stress. As a consequence of these factors attention has turned to the potential needs for higher standards of regulation for nitrate and pesticides to protect drinking water supplies, combined with further public awareness-raising and education.

### The Hunze Valley, Drenthe Province, Netherlands

The experience in the Hunze Valley illustrates the potential for spatial planning on a landscape scale to achieve multiple benefits. Visioning at landscape scale has been important and foresees 'win–win' scenarios in which, for example, sustained abstraction of drinking water occurs from a restored wetland and nature reserve that protects groundwater quality and reduces downstream flood risk, and tourism and local economic development are stimulated, helping to diversify rural incomes and reduce reliance on intensive farming, while sustaining the incentive for local environmental protection.

Over time the local authority (Province of Drenthe) has accumulated experience in defining the technical criteria to be applied in such spatial planning and in methods of public engagement. This experience shows that although there is potential for 'win–win' scenarios, there are also inevitable conflicts of interest and the need for compromises. First and foremost, the whole process depends on establishing legitimacy through broad-based public acceptance. It is essential to take the opinions and interests of all stakeholders into consideration, and to recognize these as legitimate. This does not mean that all interests can be fully satisfied, but gaining and maintaining public support for a plan and its implementation requires that each significant stakeholder shares a common understanding of the plan and awareness of the legitimate interests of others. Concerns about threats to current interests must be balanced by awareness of the opportunities and benefits that a plan may present. The costs of these processes have been significant but are considered by the Province of Drenthe to have been justified by the multiple benefits generated and by ongoing public support to sustain those benefits in future. In turn, monitoring of land management changes and public dissemination of outcomes is essential to sustain public acceptance and support.

## In conclusion

The three cases in this chapter share a number of commonalities. Each has a priority concern for protection of groundwater as a source of unpolluted drinking water, and for each the primary sources of nitrate and other pollution are the same. The agencies that led action in each case also initially lacked the data and expertise to make well-founded and sustainable management decisions. Each recognized the need to build a process that would provide the means for identification of technical solutions and successful and sustained implementation.

Each learnt that the details of solutions have to be developed locally and regionally according to the conditions that prevail. There are also many differences that range from physical catchment conditions, through legal and funding arrangements, to the interests and priorities of local stakeholders and experts. Despite these differences, some further commonalities of approach, process and action can be identified that have wider applicability.

First, the objective of groundwater protection cannot be delivered by a single authority alone. It necessarily involves a variety of stakeholders with diverse legitimate interests, including state and local authorities, water suppliers, water consumers, farmers, other rural businesses and environmental organizations. The development of plans and identification of solutions must start from a process of genuine participation by these groups in which both a common vision and a common knowledge base are established. Such participation and information-sharing must be sustained as intervention progresses iteratively and adaptively, sustaining public support for and 'ownership' of the measures that are implemented. As in the Province of Drenthe in the Netherlands, the sustained success of the whole process depends on establishing legitimacy through broad-based public acceptance and support. The safety of drinking water supplies can clearly provide an effective catalyst for such collaboration and collective action. Thus water suppliers are often key organizations to facilitate and lead programmes, contributing both their technical capacity and the financial resources of their water customers.

A second key theme demonstrated by these three cases is that programmes and policies with a single focus such as the reduction of nitrate pollution are likely to be less effective. In contrast, an integrated and holistic approach aiming for multiple benefits and capable of resolving trade-offs is both achievable and a key element in gaining stakeholder support and additional financial resources. Plans to protect groundwater should be integrated with plans for other land management goals such as recreation, landscape amenity, biodiversity conservation, and a growing and sustainable rural economy that provides livelihoods for local communities. Thus each case in this chapter sought to combine interventions to exploit synergies and achieve multiple benefits where possible. Solutions to diffuse water pollution can deliver additional benefits, and these synergies provide opportunities to combine funding streams and achieve horizontal coordination of otherwise sector-specific action plans and interventions. Again, what follows is that all stakeholders and all relevant agencies or organizations must be adequately consulted and involved in the decision-making process.

Thirdly, planning and action in each of these three cases was underpinned by sound technical assessments that provided a conceptual and practical understanding of the local and regional groundwater system in the context of its soils, hydrogeology and water body characteristics. This technical knowledge underpinned identification and implementation of solutions from field-scale farm management practices to integrated landscape-scale spatial plans, but only when the information was widely available, shared and trusted.

Finally, each case demonstrated that cost-effective solutions require a combination of measures including regulation, persuasion, voluntary action and financial incentives. In each case regulation alone was judged incapable and inequitable as a means to deliver the necessary commitment and changes in behaviour on the part of land users and local residents.

## Acknowledgements

This chapter draws extensively on the experience and reports of the Water4all and WaterCost projects (funded by the European Regional Development Fund's INTERREG programme), and presentations by, discussion with, and field visits hosted by leading participants in those projects from the three cases described here. Special thanks are due to Gitte Ramhøj, Christina Aue, Nico van der Moot, Per Gronvald, Auke Kooistra, Lars Delfs Mortensen and Onno Seitz for these contributions.

## Note

1   For maps and further description see: http://planninglaw2009.land.aau.dk/doc/slides_drastrup.pdf.

## References

Alliance Environnement (2007) *Evaluation of the Application of Cross Compliance as Foreseen under Regulation 1782/2003*, Institute for European Environmental Policy (IEEP) and Oréade-Brèche Sarl, London.
Aue, C. and Klassen, K. (2005) *Water4All, Sustainable Groundwater Management: Handbook of Best Practice to Reduce Agricultural Impacts on Groundwater Quality*, A-K Print, Aalborg Kommune, Oldenburg and Aalborg.
Brouwer, F., Heinz, I. and Zabel, T. (eds) (2003) *Governance of Water-related Conflicts in Agriculture – New Directions in Agri-environmental and Water Policies in the EU*, Kluwer Academic Publishers, Dordrecht/Boston.
Council of the European Communities (1991a) *Directive Concerning the Placing of Plant Protection Products on the Market (91/414/EEC)*, Official Journal of the European Communities, L230, Brussels.

Council of the European Communities (1991b) *Directive Concerning the Protection of Waters against Pollution Caused by Nitrates from Agricultural Sources (91/676/EEC)*, Official Journal of the European Communities, L375, Brussels.

Council of the European Communities (1998) *Directive Relating to the Quality of Water Intended for Human Consumption (98/83/EC)*, Official Journal of the European Communities, L330, Brussels.

Council of the European Communities (2000) *Directive Establishing a Framework for Community Action in the Field of Water Policy (2000/60/EC)*, Official Journal of the European Communities, L327, Brussels.

Council of the European Union (2006) *Directive on the Protection of Groundwater against Pollution and Deterioration (2006/118/EC)*, Official Journal of the European Union, L372/19, Brussels.

European Commission (2013a) *Legislation: The Directive Overview*, www.ec.europa.eu/environment/water/water-drink/legislation_en.html, accessed 6 September 2013.

European Commission (2013b) *The Nitrates Directive*, www.ec.europa.eu/environment/water/water-nitrates/, accessed 6 September 2013.

Heinz, I. (2008) 'Co-operative agreements and the EU Water Framework Directive in conjunction with the Common Agricultural Policy', *Hydrology and Earth System Sciences*, vol. 12, no. 3, pp. 715–726.

Madsen, L. M. (2001) 'Location of farm woodlands in Denmark: a quantification of the results of the scheme for field afforestation in Ribe and Vejle counties', *Geografisk Tidsskrift, Danish Journal of Geography*, vol. 101, pp. 87–100.

WaterCost (2008) *WaterCost: Elements of Cost-effectiveness Analysis*, North Denmark EU-Office, Stormtryk, Aalborg.

# 8   The WWF RIPPLE project (Rivers Involving People, Places and Leading by Example), Ulster, Northern Ireland

*Alex Inman and Mark Horton*

## Introduction

This chapter describes the development and implementation of a community-led catchment plan in Northern Ireland. The work was led by an environmental charity and delivered largely by volunteer catchment champions. For this book, this case study provides illustrations of the possible strengths and weaknesses of a community-led and locally-based catchment planning process.

## Context and drivers

### Key characteristics of the catchment

The RIPPLE project encompasses the Ballinderry River catchment located in the region of mid–Ulster, Northern Ireland. The catchment covers an area of 487 square kilometres, mostly in County Tyrone; however, the lower river, below the village of Coagh, forms the boundary between Counties Tyrone and Londonderry (also known as County Derry). The Ballinderry River rises in a small lough called Camlough, nestled between the peaks of Evishanoran and Craignagore mountains, on the southern slopes of the Sperrin mountain range. At 211 metres above sea level, the river, a small streamlet at this stage, leaves Camlough and begins a 47-kilometre journey, at first flowing northward for a few kilometres before turning east towards Cookstown and then to its mouth on the western shore of Lough Neagh; the largest freshwater lake in the British Isles.[1] Joining the main Ballinderry River, along its course, are many small streams which carry water from the surrounding countryside, and a number of tributary rivers, such as the Lissan Water, the Ballymully River, the Rock River, the Claggan River and the Killymoon River; each significant rivers in their own right (BREA, 2010).

The Ballinderry is characterised as a typical fast-flowing mesotrophic river. The upper river is dominated by bryophyte (mosses and liverworts) plant communities. The middle and lower river flows through a more managed agricultural landscape and the plant communities are less diverse. The riverbank plant communities vary along the length of the river, though wooded pockets are common, including wet woodland and some ancient woodland still present near Cookstown and at Loughry on the Killymoon River. The rare orchid Marsh Helleborine (*Epipactis palustris*) has been recorded along the Ballinderry. The river is one of only five rivers in Northern Ireland known to still hold a population of the globally endangered freshwater pearl mussel (*Margaritifera margaritifera*), though its numbers are low and may be fewer than a thousand. In addition, the river is home to otter (*Lutra lutra*), kingfisher (*Alcedo atthis*), dipper (*Cinclus cinclus*), the globally threatened white-clawed crayfish (*Austropotamobius pallipes*), brook lamprey (*Lampetra planeri*), Atlantic salmon (*Salmo salar*) and river brown trout (*Salmo trutta*). The river is also home to the unique Dollaghan trout, which is found only in Lough Neagh and its feeder rivers. It is because of these fish that the Ballinderry is often referred to as one of the best angling rivers in Ulster (BREA, 2010).

From near its source at Camlough to just above the town of Cookstown, the Ballinderry River has been given national and international special protection status because of the rare and special plants and animals which live in it. In the year 2000, the upper part of Ballinderry River, from near its source at Camlough to the Glanavon Weir at Cookstown, was recommended to the European Commission as a candidate Special Area of Conservation (SAC), because it supports a number of the rare or threatened species and habitats listed in Annexes I and II of The European Commission Directive on the Conservation of Natural Habitats and of Wild Fauna and Flora (92/43/EEC), more commonly referred to as the 'Habitats Directive'. The designated area, known as the Upper Ballinderry River SAC, was fully adopted in 2005 because of the presence of the globally endangered freshwater pearl mussel, as well as substantial otter numbers and beds of water-crowfoot (*Ranunculus spp*). In 2000, the same part of the river designated as a SAC was declared by the now Northern Ireland Environment Agency (NIEA) as an Area of Special Scientific Interest, under Article 14 of the Nature Conservation and Amenity Lands (Northern Ireland) Order 1985 (BREA, 2010). In addition, the lower part of the Ballinderry River, from the village of Coagh to Lough Neagh, falls within an area known as the Lough Neagh Wetlands, which is designated as a Ramsar site. This area of the catchment is also recognised as an internationally important area for migratory wildfowl.

### Water management challenges and their severity

The Ballinderry River faces many pressures including agricultural pollution, urban and industrial pollution, barriers to fish migration, underperforming or inadequate sewage treatment works, excessive water abstraction, pollution from landfills, quarries, contaminated land and the negative impacts of invasive alien species. As with many catchments across Northern Ireland, the cumulative effect of poorly managed septic tanks pollutes the streams that feed the main river. Acute agricultural and industrial pollution is rare, but when it happens it can have a significant impact on freshwater invertebrates, fish and the wider ecology of the river. One of the biggest problems is siltation. Excessive amounts of silt washed out from eroding banks – often trampled by cattle – and from quarry operations in the catchment, clog up the gravel beds in which trout and salmon lay their eggs. As aerated water cannot get through the silt to the eggs, the eggs suffocate and die. The freshwater pearl mussel also depends on clean water and well-aerated gravel to survive. Unfortunately this species is facing extinction in the Ballinderry unless extensive water quality and habitat improvements are made.

### Drivers and timeline of the RIPPLE project

The RIPPLE project did not emerge from a standing start but from a position of considerable momentum, facilitated by the social capital already existing within the catchment as a result of partnership work between WWF (World Wide Fund for Nature) and a charity named the Ballinderry River Enhancement Association (BREA).

WWF is a well-known international environmental non-governmental organisation (NGO) which primarily works at a policy level on a wide range of conservation and natural resource management issues. In 2002, WWF received financial support through a partnership with the HSBC Bank which became focused on a UK-wide freshwater resource management programme. The UK Natural Rivers Programme had a number of different themes but in Northern Ireland the focus was on water quality improvement and pollution mitigation. Working with the Ballinderry Fish Hatchery Ltd (BFH), a not-for-profit community business which provides advisory, fish breeding and river restoration services and is owned by its parent charity the BREA, WWF launched a two-year project in 2005. This was the Ballinderry River Enhancement Project which concentrated on diffuse pollution from agriculture and set out to demonstrate best land management practice. Under this project farmers were encouraged to improve river water quality and habitats through participation in government funded agri-environment schemes. Public education events were also held and a video entitled 'Bridging Troubled Water' was produced. Distributed in DVD format, the video presented simple, low-cost steps that farmers can take to protect Northern Ireland's rivers and loughs, helping them to meet national and EU water quality legislation. These measures quickly led to a significant improvement in water quality in a Ballinderry tributary where efforts had been concentrated. Finally, management plans for the river, its habitats and species were produced.

WWF then funded a further one-year project to 'kick-start' a community-led river basin management plan designed to parallel the official NIEA river basin management plans being developed to meet the statutory requirements of the EU Water Framework Directive (WFD). WWF has been critical of the way UK government agencies have implemented the WFD, particularly with regard to public participation. By funding a community-led process WWF wished to demonstrate best practice in participatory catchment management planning as a benchmark against which to assess the NIEA-led process.

This community initiative (2007–2008), 'Stepping Stones to RIPPLE', formed the basis for the RIPPLE project, which WWF contracted the Ballinderry Fish Hatchery Ltd to deliver. Financial support from the Heritage Lottery Fund (HLF; a charitable grant provider) and match funding provided by WWF funded the RIPPLE project for three years from 2009 to 2011. Funding for improvement of the Ballinderry River has therefore been iterative in nature and achieved through a series of initiatives and projects.

### Organisation, legal status and wider governance arrangements

Established in 1984, BREA is an 'umbrella' organisation for river interest groups on the Ballinderry River aiming to 'restore the Ballinderry River to its former greatness'. It has a catchment-wide focus with a remit to improve water quality, restore habitats, protect species and improve recreation along the river and its tributaries. Educating the community 'on the importance of the river environment' is also a key stated objective.

The association has charitable status, as recognised for tax purposes in Northern Ireland by HM Revenue and Customs, and is led by a group of voluntary committee members and a voluntary board of trustees. In common with many small environmental NGOs, funding is sporadic and generated from a variety of grant-giving sources. In addition, Ballinderry Fish Hatchery Ltd provides a modest income stream to help fund habitat enhancement projects, restocking programmes and educational initiatives in the Ballinderry River system. In 2006, BREA became a member of The Rivers Trust (www. theriverstrust.org), a national organisation for river conservation charities and groups across England, Wales, Northern Ireland and the Republic of Ireland.

Over the last 25 years BREA has worked at a local level undertaking hands-on habitat restoration works and restocking programmes on the river. It has also run an active and ongoing education programme with local schools and undertaken outreach work, as well as hosting visits at a bespoke education facility built at the Ballinderry Fish Hatchery. This provides visitors with the opportunity to get up close with native fish, freshwater pearl mussels and white-clawed crayfish at various stages in their life-cycle. To date the activities of BREA have thus been very applied and practical in nature with a focus on habitat and species rehabilitation, and a particular emphasis on fisheries management.

As in many areas of the European Union, the wider governance and management of water resources in Northern Ireland is presided over by a number of government departments and statutory agencies. The key entities in the governance arrangements for the Ballinderry catchment are described below.

The Department of the Environment (DOE) is responsible for the protection of the aquatic environment through the regulation of water quality, and the conservation of freshwater, marine flora, fauna and hydrological processes. The implementation of legislation and environmental monitoring is undertaken by DOE's executive agency, the NIEA.

The Water Management Unit (WMU) within the NIEA has a duty to promote the conservation of the water resources of Northern Ireland and the cleanliness of water in waterways and underground. It undertakes a number of activities including monitoring water quality, preparing water quality management plans, controlling abstraction and effluent discharges, taking action to combat or minimise the effects of pollution and supporting environmental research. The NIEA's Biodiversity Unit is responsible for protecting Northern Ireland's biodiversity, including freshwater biodiversity through species and habitat action plans. Designation and management of Areas of Special Scientific Interest (ASSI) and Special Areas of Conservation (SAC) fall to another unit called Conservation Designations and Protection (CDP).

The Inland Fisheries Division of the Department of Culture, Arts and Leisure (DCAL) provides advice and guidance on matters relating to the conservation, protection, development and improvement of salmon and inland fisheries to angling clubs, fishery owners and a range of other water users and interested parties. These powers are discharged under the provisions of the Fisheries Act (NI) 1966, for the inland fisheries of Northern Ireland.

Through its Fisheries Division, the Department of Agriculture and Rural Development (DARD) is responsible for sea fisheries, aquaculture and fish health policy; the enforcement of fisheries legislation; the licensing of aquaculture; fishing vessel licensing; and the administration of fisheries grant schemes. The DARD is also the statutory drainage and flood protection authority for Northern Ireland through its Rivers Agency. Under the terms of the Drainage (Northern Ireland) Order 1973 the Rivers Agency has discretionary powers to maintain watercourses and sea defences which have been designated by the Drainage Council for Northern Ireland, to construct and maintain drainage and flood defence structures and to administer advisory and enforcement procedures to protect the drainage function of all watercourses. All executive functions arising from the department's statutory remit under the Drainage Order are undertaken by the agency and it also exercises the department's responsibilities in regard to regulation of the water levels in Lough Neagh and Lough Erne (the latter in conjunction with the Electricity Supply Board in the Republic of Ireland).

The Department for Regional Development (DRD) in addition to having responsibility for transport and regional strategic planning undertakes policy and support work for air and sea ports and oversees policy on water and sewerage services and management of the department's shareholder interest in Northern Ireland Water.

The latter is a government-owned company, that is a statutory trading body owned by central government but operating under company legislation and with substantial independence from government, which provides water supply and wastewater services across Northern Ireland.

Lastly, the Council for Nature Conservation and the Countryside was established in 1989 under the provisions of the Nature Conservation and Amenity Lands (Amendment) (Northern Ireland) Order 1989. Its statutory role is also covered by the provisions of The Environment (Northern Ireland) Order 2002. The council is made up of 16 members who cover a wide range of environmental expertise and experience throughout Northern Ireland. Its role as statutory advisor is to advise the DOE on matters affecting nature conservation and the countryside. It also offers advice relevant to its remit to other government departments such as the DARD, DRD and DCAL and is represented on other groups and working parties.

Faced with this institutional complexity in the public sector, WWF and BFH were keen from the outset to build strong working relationships between members of the local community engaging with the RIPPLE project and the technical agencies and decision-making bodies responsible for water resources management within the Ballindery catchment. To this end, BFH facilitated the establishment of an advisory group to act as a point of contact between decision-makers and members of the local community. This group had no formal terms of reference and met infrequently. Its function was to serve as a list of individual people who could be called upon on an ad-hoc basis to provide advice and assistance when needed, rather than have any decision-making capacity within the project. Members of the advisory group are listed below.

- Cookstown District Council;
- Cookstown and Western Shores Area Network (CWSAN; a community umbrella organisation formed in 1996 to represent and support the interests of community groups within the rural areas of Cookstown district and along the western shores of Lough Neagh);
- Council for Nature Conservation and the Countryside;
- Department of Agriculture and Rural Development – Countryside Management Branch;
- Department of Culture, Arts and Leisure – Inland Fisheries;
- Department of Agriculture and Rural Development – Rivers Agency;
- Department of the Environment – Environmental Policy Division;
- Department of the Environment – Planning Service;
- Northern Ireland Environment Agency – Biodiversity Unit;
- Northern Ireland Environment Agency – Conservation Designation and Protection;
- Northern Ireland Environment Agency – Water Management Unit;
- Lough Neagh and Lower Bann Advisory Group;
- Northern Ireland Water;
- Rural Community Network.

Since December 2003, river basin planning in the UK has been shaped by the EU Water Framework Directive, established in law in Northern Ireland via the Water Environment (WFD) Regulations 2003 (SR 2003 No. 544). These regulations identified the Department of the Environment as the competent authority to coordinate the implementation of the directive with its executive agency, the NIEA, leading all technical aspects pertinent to WFD implementation. The Department of the Environment is working in partnership with the Department of Agriculture and Rural Development, the Department of Culture, Arts and Leisure and the Department for Regional Development to implement the directive. To oversee and coordinate implementation, an inter-departmental board has been established with an implementation working group coordinating the activities of the government departments and agencies responsible for delivery.

In keeping with the requirements of the directive, the Department of the Environment introduced a six-yearly cycle of planning, action and review. Three river basin management plans were initially produced by the NIEA, for three of the four river basin districts[2] in Northern Ireland. The fourth river basin management plan was produced by the Republic of Ireland's Department for Environment, Community and Local Government, as only a small portion of this river basin district lies within Northern Ireland. The Ballinderry catchment resides within the Neagh Bann River Basin District, which is one of the International River Basin Districts designated within Northern Ireland.

In contrast to the Neagh Bann international river basin management planning process, the RIPPLE project was, from the start, an unofficial catchment initiative with no formal legal standing and no statutory remit. It was funded independently from governmental financial mechanisms, run by independent locally based environmental NGOs with volunteer assistance, and had no statutory delivery targets or reporting requirements.

Despite this lack of official standing, it does not appear that the existing governance hierarchy stood in the way of the RIPPLE project. Indeed, as outlined later in this chapter, the statutory agencies were supportive of the initiative and invested time through membership of a project advisory group. A later development was a mapping exercise between the NIEA and WWF/BFH to determine how the action plan developed within RIPPLE could assist in the delivery of the WFD river basin management plan. Given that the RIPPLE project complemented many elements of the WFD plan, it was not surprising that the NIEA was keen to lend its support. This complementary relationship was perhaps a key reason why the unofficial status of RIPPLE was not an impediment to its successful delivery.

It is also noteworthy that while it lacked official standing, the RIPPLE project gained moral standing and legitimacy within the local community, a position which was recognised by the statutory agencies and local government entities. This led to additional financial support – for example, Cookstown District Council provided financial assistance to support certain outputs from the project.

# Getting started

## *Leadership (people and organisations)*

Many people contributed to the RIPPLE project but the core team consisted of the following. The project was managed by WWF Northern Ireland Freshwater Policy Officer, Dr Claire Cockerill. Claire oversaw the delivery of project outputs within budget, ensured the commitments to funders were met and promoted the policy outputs from the project. Dissemination of lessons learned from the project occurs across Northern Ireland, the UK and the EU. Christine Crawford, WWF Northern Ireland's communications manager, was responsible for the delivery of the communications elements of the project, including production of a newsletter three times a year, regular press columns, website and podcasts, and promotion and organisation of project events. The project was coordinated on location by the Ballinderry Fish Hatchery Ltd and in particular by their coordinator of new projects, Mark Horton. As experience in partici-patory catchment management processes was initially lacking, it was recognised that additional expertise would help a project based on participatory processes, and the assistance of the Rural Community Network (RCN) was engaged.

Based in the local town of Cookstown, the RCN is a regional voluntary organisation established by community groups from rural areas in 1991 to articulate the voice of rural communities on issues relating to poverty, disadvantage and equality. It is a membership organisation and is managed by a voluntary board of directors, elected every two years. Two community representatives from each of the six counties in Northern Ireland make up the main component of the Network's board, with statutory, voluntary, farming, environmental, cross-border and other sector groups making up the rest.

## *Processes of partnership building and community engagement*

A 'cornerstone' of the RCN's approach is the use of community development[3] and networking to empower communities to address poverty and social exclusion. In particular, the RCN champions the role of volunteers in building social cohesion and community capacity. Action research undertaken by the RCN had previously highlighted the potential of using people's own interests, and in particular those related to the environment, as a trigger for engaging them in community development activities. It was also learnt that a community development initiative needs to employ methods and models which are appealing and accessible to people who may not be interested in formally constituted community groups. Research and experience showed that an intervention based on community development should first establish in conjunction with the community what the social, physical and human assets are that can be built upon to stimulate and sustain activism.

The shape of the RIPPLE project became strongly influenced by the RCN's emphasis on community development and an assets approach. Indeed, it is possible to characterise the RIPPLE project as a community development project rather than as a natural resource management project. It became a project in which the Ballinderry River was used as a focus around which to stimulate community action, rather than as an environmental resource to be protected for its own sake. Core values of community development which underpin its practice are: collective action (promoting the active participation of people within communities), community empowerment (promoting the rights of communities to define themselves, their priorities and agendas for action) and working and learning together (enabling communities to learn from reflecting on their experiences) (Life Long Learning UK, 2009).

An ongoing ethos in the RIPPLE project was to maintain a balance between community empowerment and science-led decision-making. This was achieved through a transparent planning phase where both community-inspired and science-based agendas were accommodated within an open discussion forum. Giving equal time and emphasis to the voices of local community members and scientific professionals created a constructive dialogue in which members of the community could feel their opinions were taken seriously. Where the community suggested initiatives with the potential to have negative impacts on the river system, these initiatives were usually voluntarily withdrawn following the application of a shared learning approach, which enabled the community to gain an appreciation of the likely consequences of the proposals put forward.

Throughout the project, BFH assumed the role of a bridging agent or neutral broker, investing significant time and energies in developing trust with and between project participants. Generating an atmosphere of collective responsibility was a key aim. For example, when addressing the issue of water quality, BFH has been very careful not to isolate any particular sector as the cause of the problem, rather emphasising that all sectors contribute to the problem and all therefore have a responsibility to take remedial action. This has gone a long way to gaining the trust and engagement of the farming community who have traditionally been singled out within water resources management plans as the sole culprits for water pollution in rural areas.

Personal testimonies from community participants attending initial RIPPLE meetings suggest they were often 'high-tension' events, particularly over the issue of access to, and recreational use of, the river; with the chief protagonists usually being anglers, canoeists and members of the farming community. What is interesting from these testimonies is that these tensions appear to have been almost entirely resolved through facilitated dialogue, providing the opportunities for the parties to explore solutions to their differences in a non-combatant atmosphere. Individuals taking part in the process were specifically asked to engage with the project with a positive problem-solving mindset, and to refrain from reverting to entrenched positioning. In particular, participants were asked to look forward rather than backwards to encourage ways of circumventing historic sources of conflict.

I remember some of the initial meetings being very high-tension events, with long-standing differences of opinion being expressed. However, these tensions seemed to get resolved because everything was taken seriously, written down and discussed. Somehow, people were encouraged to sort out their differences and were made to work together for their tea and buns.

(RIPPLE champion)

Where access to land has remained an area of tension, BFH has continued to build trust between conflicting parties through ongoing dialogue and by engaging appropriate third parties where necessary. For example, to allay the fears of a landowner near Cookstown concerning liability risk associated with greater public access across his land, BREA contacted another landowner with established public access arrangements and presented this case study as a working model. The result was the opening up of the land near Cookstown via an informal agreement between the landowner and the BREA with no legal or administrative costs incurred.

Research has demonstrated (e.g. Fell *et al.*, 2009) that individual personalities can be extremely influential in the building of trust across social networks. The RIPPLE project would appear to offer a good example. It is very clear that people across all stakeholder groups involved with the RIPPLE project came to respect and trust the project coordinator. As described by a RIPPLE volunteer:

I see him as someone with no agenda who believes in what he is doing. He just wants to help the river and has the community's best interests at heart. He gets a lot out of people by not being pushy and being very diplomatic.

## Approach and tools

Once embarked on community-led participatory planning the project team set about designing a carefully thought out stakeholder engagement process, involving multiple stages and multiple actors. What is noticeable from an examination of this process is the length of time taken from project inception to the arrival at an agreed catchment management plan. This was a period of 24 months. This timeline reflects the high level of interaction which took place with the community and the iterative nature of the methodologies used. A schematic of the key stages in the process is outlined in Figure 8.1, and followed by a narrative of the process design and its implementation.

The RIPPLE project and its preceding 'stepping stones' initiative generated both a strategic vision and the mobilisation of a practical action plan involving on-the-ground delivery. All of these activities have been the product of ongoing community involvement.

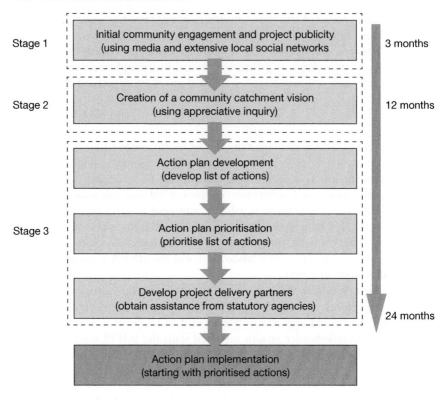

*Figure 8.1* Key stages in 'stepping stones' to RIPPLE and RIPPLE project catchment plan development (Source: Authors)

### Stage 1: Initial community engagement and project publicity (three-month duration) under the 'stepping stones' to RIPPLE phase

Given that the decision had been made to use a community development approach, significant effort was channelled to publicise the project to as broad an audience as possible, and not just interest groups traditionally associated with water resource projects such as angling clubs and environmental action groups. A wide range of communications media were used including local press, radio and poster campaigns and local community news bulletins. A key message in all such communications was that the project was open to all people with an interest in their local community resources, of which the Ballinderry River is a part. Also key to the project's initial engagement strategy was a launch event held at a venue of particular cultural significance to the community – Lissan House, situated two miles north of Cookstown – that was previously inaccessible. Providing the local community with an opportunity to gain access to the property for the first time ensured RIPPLE attracted a wide range of interested parties including youth groups, local history clubs, members of the artistic community and local community associations; none of which had existing links with river restoration activities. The event was designed as an opportunity for the local community to explore and celebrate the social and cultural significance of the Ballinderry River and the role the river has played in local music, poetry and storytelling. WWF and BFH made extensive use of existing social networks to publicise the Lissan House event. A partnership was formed with a local community support network – the Cookstown and Western Shores Area Network – to again help target a range of interest groups outside angling and conservation interests. Inviting a well-known broadcasting celebrity, Joe Mahon, to speak at the event also added broader appeal and interest.

The success of this and subsequent events in attracting a broad profile of community members is evidenced by the variety of organisations that have actively engaged with the project since its inception. A selection of the participating entities not usually associated with a water resources project of this nature are listed below (the commitment of a large number of environmental and angling organisations is not listed as their involvement is not atypical in projects of this nature). In total, 85 separate organisations and 256 individuals have participated to varying degrees in the development of the project.

- Ardboe and Coalisland Knitting and Crochet Groups;
- Ballinderry Community Group;
- Ballinderry Historical Society;
- Ballinderry Rural Development Association;
- Coagh Historical Group;
- Cookstown History Group;
- Cookstown Scouts;
- Cookstown Young Farmers;
- Countryside Access and Activities Network;
- Coyles Cottage Women's Group;
- Friends of Duncairn;
- Friends of Lissan Trust;
- Girl Guides;
- Granard Community Group;
- Kinturk Cultural Association;
- Loup Women's Group;
- Mid-Ulster Bee Keepers' Association;
- Mid-Ulster Photography Club;
- Moneymore Community Group;
- Moneymore Heritage Trust;
- Moneymore Young Farmers;
- Muintirevlin Historical Society;
- Muntervick History Group;
- Orritor Primary School;
- Positive Steps Women's Group;
- Probus Club Cookstown;
- Stewartstown and District Local History Society;
- Wellbrook Beetling Mill.

### Stage 2: Creation of a community catchment vision for the river (nine-month duration) under the 'stepping stones' to RIPPLE phase

After the initial publicity stage of the project, a series of seven evening meetings were held at various locations within the Ballinderry catchment to begin the process of building a shared vision for the river. These meetings were arranged at community venues to ensure open and easy access and were structured to last one and a half hours each. Seven meetings across the catchment ensured members of the community did not have to travel far to attend, which facilitated their involvement with the project. It was also the view of WWF and BFH that communities within the catchment would initially resonate more with their local tributary or section of the river, rather than the Ballinderry catchment as a whole. Subsequent experience generated from the meetings showed this hypothesis to be correct, although participants began to take a more holistic view of the catchment the longer they were involved with the project.

Meeting invitations and subsequent introductions at the meetings focused on the value of local community experiences and knowledge of the river. Meeting participants were informed that their input would be used in the production of an action plan, and to manage expectation it was made explicitly clear which areas of future management they could and, importantly, could not influence. To ensure the process helped develop participants' knowledge base of the policy context within which the RIPPLE project was operating, an overview of the Water Framework Directive was presented. In particular, the main aims of the directive were introduced, why the directive was relevant to the local community and how the community could get involved in the WFD consultation process.

Participants were asked to consider how the river might be used as a resource for the community and, as such, how the river might be nurtured as a community asset. Other community assets were identified including the community itself. Appreciative inquiry[4] was used to identify these assets and build subsequent actions by considering the best of the past and the present and then creating a vision for the future. Memories were explored regarding human interactions with the river and any particularly positive recollections of these interactions. This set the scene for a visioning exercise. Participants were provided with pictures, magazines, pens and prompt words and were asked to work in breakout groups to create a visual representation of their vision for the Ballinderry. By basing these visions on past and present realities established through appreciative inquiry, it was possible to achieve realistic goals and manage participant expectations of what might be achieved in the future.

Following the sharing of the visions, participants were asked to consider how the actions might become a reality, who should be involved in the process and how might they be involved. In terms of involvement, participants were specifically asked to consider whether they wished to get involved in an 'action group' or whether they were happier to keep in touch with the project rather than take an active implementation role. From the 120 people who attended the events, 20 per cent agreed to join an action group. At this stage, the project team along with members of the action group began to identify which agencies and external organisations might be able to assist in the implementation of the developing management plan.

It had been noted at the visioning meetings that certain groups were under-represented, most noticeably young people, women, farmers and the business community. The project team subsequently targeted organisations representing these members of the community and ran bespoke visioning events for each group in turn to ensure their voice was heard.

All comments and suggestions from the visioning events were then grouped together by the core project team and presented at an event. This provided an opportunity for those taking part in the original visioning events (by then totalling 200 people and formalised under the banner of a Catchment Management Forum[5]) to meet face-to-face to critically examine the information that had been collated during the process thus far. Those people unable to attend the event were sent the material by email or hard copy for their comment.

Feedback from this stage of the process was then used to refine the visions into a draft vision document which identified the following four priorities:

1   better access along the river;
2   better understanding of the river environment;
3   a cleaner river;
4   more wildlife.

This draft document was presented to and subsequently ratified by the Catchment Management Forum (WWF, 2008). What is important about the priorities listed in the document is that they were community led, not driven by specific legislative drivers, and therefore represented the aspirations of the local community. This is important because it is possible to detect significant ownership of the resulting action plan (see below) by the community, an ownership which is often lacking where priorities are perceived to be imposed by an external agency. In contrast to the Water Framework Directive planning process which is dominated by water quality and ecological priorities, the RIPPLE project enabled the community to incorporate river access and education priorities, which are clearly salient issues for many local people in the Ballinderry catchment. The experience clearly showed that the opportunity to include access and education priorities engaged a broader spectrum of people than would have been the case had water quality alone been the sole focus for the project.

### Stage 3: Action plan development and prioritisation of actions (twelve-month duration) under the RIPPLE project

Following wide acceptance of the visioning document, participants who had expressed an interested in joining an action group were invited to attend a series of action planning meetings. The purpose of these meetings was firstly to record all actions considered necessary to make the visions a reality, and secondly to prioritise these actions using an 'Importance/Achievability' grid. In total, the community proposed 115 actions for realising their visions. Following the prioritisation process, 29 actions across the four priority aims identified in the vision document were highlighted for development during the first year of the implementation phase. Examples of these actions are presented in Table 8.1.

*Table 8.1* Examples of actions for the first year of Action Plan implementation, RIPPLE Project

| Better access | Better understanding | Cleaner river | More wildlife |
|---|---|---|---|
| Remove all invasive alien plant species from access routes to make them safe for use | Provide training days for local community on how to spot pollution | Identify the septic tank discharge carrying capacity of each watercourse in the catchment | Report the locations of all known non-native invasive plants and animals in the river system |
| Clarify liability and access rights issues for landowners providing access and access users | Organise and promote community 'river days' to raise awareness of the river | Promote the safe use and disposal of household chemicals and cleaning products | Extend the range of the freshwater pearl mussel |
| Work with landowners to agree new access | Encourage schools in the river system to use the river as an outdoor classroom | Raise awareness of the impact dangerous industrial chemicals can have on water quality and wildlife | Collect all known information on river wildlife and habitats in the Ballinderry River system |
| Encourage more young people to use the river | Add information boards to access routes on themes such as wildlife, local history and river facts | Clarify responsibility for removing fallen animals from watercourses | Establish a community based non-native invasive plant eradication programme |

In contrast to conventional catchment planning in the UK these prioritised actions were framed entirely from the aspirations of the community and were not governed directly by legislative drivers or conservation targets. Where relevant to assist with planning, participants were provided with technical guidance by BFH or members of the advisory group. For example, BFH advised that a hogweed eradication programme would need to be preceded by a comprehensive hogweed mapping exercise, a recommendation which was adopted by the community. For each action a process was followed which involved decisions on:

1    steps that should be taken to get the action started;
2    a list of those people and organisations that may need to be involved in delivering the action;
3    a timeframe for delivery;
4    the measurable outcome;
5    who in the community would lead the delivery of the action.

At this stage thematic action sub-groups were formed to take specific actions forward, each sub-group coordinated by a volunteer champion to manage the activities and develop them into deliverable projects. The notion of champions was conceived by the action group itself and not by BFH, one indicator of the level of self-determination that RIPPLE volunteers had by this time adopted. This process culminated in the development of the 'Ballinderry River Action Plan – Phase 1' which was ratified by the Catchment Management Forum through a further process of consultation and amendment.

Having determined a draft action plan for each theme, each champion, with the support of the project coordinator, decided which agencies and external organisations should be engaged to help facilitate project delivery. This process culminated in a 'speed-dating' event between the champions and members of the project advisory group. The project team facilitated this event by managing pre-meeting expectations and allaying initial fears from the advisory group that the champions would present too many unrealistic demands. At the event, champions rotated around the various service providers present to secure commitments via 'promise slips' which were mapped on a display board. Most commitments by the service providers initially involved setting up subsequent meetings with relevant members of staff or departments, which generated frustration among some of the champions. However, this process did enable relationships to begin developing between the various actors, and this has led to effective lines of communication between many of those involved.

The final action plan (WWF, 2009) was launched just prior to the Cookstown 'Banks of the Ballinderry Fair' in May 2009. An event at the Wellbrook Beetling Mill drew together members of the community, politicians, representatives of the local council, government departments and other key stakeholders to acknowledge this achievement and publicise the production of the action plan. Building on the format of previous events, educational activities and information stands were themed on the visions set out in the RIPPLE project.

### Knowledge exchange and the use of science

The RIPPLE project developed as a community-led project rather than a science-led initiative, but scientific information and expertise was sought as and when needed to help the champions and delivery groups in planning and implementing actions. A particular feature was the degree to which community volunteers took responsibility for obtaining scientific data themselves, either directly or by asking the RIPPLE project coordinator to obtain data on their behalf. This process was initially facilitated through the 'speed-dating' exercise between action champions and the project advisory group but became continuous. For example, champions requested information pertaining to the carrying capacity of streams downstream of sewage treatment works and to the maintenance of septic tanks from the NIEA. Another champion made contact with organisations including the National Museum of Northern Ireland, Queens University, the University of Ulster and the Centre for Environmental Data and Recording to obtain data on species, habitats and water quality relevant to several actions in the Ballinderry River Action Plan.

BFH was also instrumental in accessing specific technical information to facilitate plan delivery. For example, to assist RIPPLE volunteers in the development of a bio-monitoring system to survey water quality, the RIPPLE project coordinator made contact and subsequently secured the services of the River Fly Partnership (RFP), an organisation specialising in the acquisition and interpretation of riverfly monitoring data. Twenty-four RIPPLE volunteers were trained by the RFP to monitor riverfly life and establish a GIS mapping system for the storage and spatial analysis of data. Such interventions by BFH enabled RIPPLE volunteers to access best available science. Another example of this is where BFH facilitated collaboration between volunteers tackling invasive plants and the NIEA. This connection resulted in the volunteer team receiving expert advice in plant dispersion and control methods culminating in City and Guilds certification in the use of pesticides within aquatic environments.

## Getting things done

### *Implementation and operation*

The resulting RIPPLE action plan has been almost entirely implemented through community action groups, each group coordinated by a champion who, in turn, is supported by the RIPPLE project coordinator. None of the actions are being delivered solely by a statutory agency, with the exception of communication of a pollution 'hotline' number through signage across the Ballinderry catchment by the NIEA. However, a legacy of the project has been identification of a limited number of actions that cannot be delivered by community members which have thus been brought to the attention of the relevant statutory bodies for action.

Fourteen champions worked on a voluntary basis without payment or even reimbursement of personal expenses. They led on subject areas identified within the action plan from fallen animal disposal and invasive weed control to public access and engagement with business. After initially working alone the champions decided to work in a group to pool knowledge, avoid overlap and provide mutual support. They also called on the assistance of local experts where available, who also volunteered time to the project when required.

The champions proved critical to the delivery and sustainability of the RIPPLE action plan. By asking members of the community to adopt each action, local ownership of the plan has been maintained. This is unlikely to have been so pronounced had an external individual or organisation been brought in to deliver the plan. Champions have been recruited from a wide variety of backgrounds and included a local environmental manager for a quarrying company, a mechanic, a retired secondary school teacher and an ex-NIEA employee. An employee from the Ballinderry Fish Hatchery also enrolled as a champion for public access based on his intimate knowledge of the land holdings and landowners within the catchment. Between champions contributions of time varied significantly from two to twenty hours each month, depending on the nature of the action and personal commitments. The energy and commitment displayed by these volunteers was significant and impressive. For example, the hogweed champion undertook invasive weed mapping exercises, managed a team of weed sprayer volunteers and negotiated access to private land for spraying.

In addition to providing hands-on effort the champions have also acted as key intermediaries with the local community, resulting in additional people becoming involved. Examples of social capital in action abound. For example, one champion is secretary of a local photography club. Through this link the club was commissioned to exhibit in Cookstown a portfolio of riverscape pictures. This brought the club members and local residents into closer contact with the project and their catchment, and some new volunteers.

The ongoing support provided by BFH is regarded by the champions as crucial to the success of the project as it makes them feel they are part of a team and not working in isolation. They also value the RIPPLE project coordinator's role as a neutral broker, providing balance to activities and limiting undue influence by any single interest group. On a practical level BFH has also managed delivery of actions considered too resource-intensive for the champions alone. This has highlighted the need within such a project for professional and paid staff to both complement and supplement champion effort, as the efforts of willing volunteers can only go so far.

### Communication

An important contributor to the project's success and profile has been sustained effort to connect the project with the wider world through, for example, cultivating media interest and relationships. WWF has supported the BFH in communication activities both locally and further afield. For example, NHK, a Japanese broadcasting corporation showcased the Ballinderry River within a feature on conservation work, and in the UK the RIPPLE project was featured on BBC Radio 4's *Open Country* programme. In addition there have been opportunities to share experiences of wider policy relevance at national and international conferences and meetings.

At local level effective communication has been instrumental in raising awareness, garnering support for the project, recruiting volunteers, fostering cooperation and sustaining interest. A range of communication tools have been utilised to reach a wide cross-section of the community within and beyond the catchment's boundaries.

- A project website provides information about the project and publishes event details and documents. To be fresh and current web content included podcasts from events such as a clean-up day on the river, a fly tying class and interviews with RIPPLE champions.
- A four-monthly newsletter was circulated to a mailing list across Northern Ireland. This included contributions from community members and

provided updates on project events, activities, achievements, seasonal river states and relevant wider issues.

- The local newspaper featured a regular RIPPLE column, which was used to report project progress or advertise project events.
- Press releases to local and national print and broadcast media, and use of social networking sites, all communicated 'RIPPLE news' to a wider audience.
- Open, inclusive and informal local events provided opportunities for face-to-face engagement and built the project's team spirit. For example, a key event, entitled 'The RIPPLE Journey' was held at the halfway point in the project to map progress, acknowledge and celebrate achievements and encourage ongoing actions.

### Sources of finance

Finance for the RIPPLE project was derived from multiple non-governmental sources. The core costs of the project, effectively the project management costs and ongoing costs associated with hosting meetings, were financed by WWF for initial project development work and by the UK Heritage Lottery Fund with match funding provided by WWF for the three-year RIPPLE project. These core funds have been supplemented by additional grants, raised by WWF, BFH and individual champions on an ad-hoc basis to deliver specific actions. The local council has been instrumental in providing funds and accessing further funds from the Department for Culture, Arts and Leisure to undertake certain public-access related infrastructure works, and they have also facilitated other funding through the landfill tax system. Other sources have included the Woodland Trust, the National Lottery and the Lottery's Awards for All programme.

As with other aspects of the project, the champions have played a major role in funding, raising money themselves for delivery of specific initiatives. A case in point is a project to involve young people in watersports. This was championed by a student who volunteered for the RIPPLE project when home from university. Assisted by the RIPPLE project coordinator she applied for funding from the Sports Relief fund to initiate one-day taster sessions to introduce young people to the sports of canoeing, kayaking, bouldering and orienteering on the Ballinderry River.

Fundraising by the champions is an aspect of community empowerment with the potential to embed confidence and capacity within the community, thereby contributing to the sustainability of the programme. However, we later return to the question of how much funding volunteers can reasonably hope to raise and whether they can do this without the continued support of the RIPPLE project core team.

## Outcomes

BFH established a detailed operating plan to monitor the implementation of the actions generated from the community visioning exercise. At the time of writing, all bar two of the 29 actions outlined in the 'Ballinderry River Action Plan – Phase 1' had been initiated and were at various stages of completion. The project was also well underway with delivering the next 30 actions identified for Phase 2. Phase 2 actions were prioritised by the community using a similar process to that employed in the initial planning phase. A Phase 3 of the project would incorporate the remaining actions. The operating plan was distributed to all key parties involved in the project thus far.

Monitoring is in the main based on process indicators rather than quantifiable outcome indicators. For example, under the 'better access' vision priority a target action was to 'work with landowners to agree new access', and this has been achieved through the creation of a new riverside walk. However no data exists, nor were there plans to collect data, regarding the use of this new walk and how people value it. Similarly under the 'clean river' vision priority there were actions to 'promote good maintenance of septic tanks' and 'promote the safe use and disposal of household chemicals and cleaning products', but again there could be no provision to monitor actual change in household behaviours, or means to relate such change to change in water quality. Monitoring is also lacking data – both quantitative and qualitative – regarding the social capital and knowledge transfer that the project has generated since its inception. For example, while the anecdotal evidence suggests a significant proportion of the local community has become reconnected with the river, the project does not have the methodologies in place to robustly assess this outcome.

There are exceptions, and some tangible outcome indicators for the RIPPLE project are being monitored. Foremost is change in water quality as baseline and follow-up chemical and biological water quality surveys using both NIEA General Quality Assessment data from statutory monitoring points, plus data from an additional 66 sites monitored by the BREA have been undertaken since the project started. Using this data, attribution of changes in water quality to RIPPLE project actions or to other external factors will depend on the extent and intensity of other measures taken within the catchment, and future analysis will need to rely on expert judgement supported by computer modelling. Other exceptions in terms of outcome indicators arise for specific actions such as the spraying and re-mapping of invasive plants. Overall there is no lack of recognition with in BREA and among the champions regarding the value of improved monitoring approaches and outcome data.

## Concluding reflections

The RIPPLE project has been chosen as a case study for this book because it displays many of the characteristics important for a successful and sustainable catchment management programme. In 2011 it remained one of the few catchment-wide programmes in the UK within which members of the local community had exercised genuine influence in decision-making and played leading roles in implementation. Other examples do exist, and have increased in number since the catchment-based approach was rolled out nationally in England and Wales from 2013.

From WWF's perspective the RIPPLE project was developed to provide an example of best practice in public participation in river basin management planning and delivery, potentially acting as a model for catchment management in the UK and Ireland. Timed to contribute to the implementation of the Water Framework Directive, the project provides an interpretation of Article 14 of the directive which requires member states to encourage the 'active involvement of all interested parties'.

Through their networks and positions on a range of policy forums WWF are continuing to disseminate the lessons learned from the RIPPLE project, which has attracted a high level of interest from policy decision-makers and other NGOs across Europe. The benefits of the approaches adopted and successful engagement of committed local community members tackle two of the most significant and widespread issues for catchment management. Firstly, the approach is believed by WWF to provide value for money in times when government budgets are constrained, by exploiting social capital, accessing additional funding and achieving tangible environmental improvements to help realise water quality targets and improve tourist revenues. Secondly, it is clear that government departments alone cannot achieve the water quality objectives set by European directives. A change in behaviour by many individuals, businesses and organisations is required. Mobilisation of the understanding, commitment and resources of the catchment community provides the means to put policy into practice.

To achieve these ends the project has demonstrated a consistently collaborative and inclusive approach by involving all relevant stakeholders within the Ballinderry catchment in assessment and planning, and in the delivery of actions. BFH has acted as a key broker and intermediary and this has improved communication and cooperation between a diverse range of organisations and agencies across the public, private and third sectors. New layers of trust and social capital have been generated between individuals and organisations not previously familiar with one another. It is unlikely, for example, that members of the Coyles Cottage Women's Group would have become actively engaged with anglers and conservationists to help restore the river had it not been for the activities of the RIPPLE project. In addition, considerable effort has been targeted at connecting decision-makers with members of the local community, with the aim of better communication and coordination between the various levels of governance. Continued turnout at meetings and the number of people actively involved with delivery of catchment improvements strongly indicate that those involved are deriving value from the process. NIEA and DOE project officers became increasingly involved in provision of advice and other technical support to RIPPLE champions engaged in practical action; a testament to a growing collaboration between local people and the statutory authorities.

The relative simplicity and focus on process indicators exhibited by the monitoring of the RIPPLE project is characteristic of many non-governmental projects financed by ad-hoc and grant-based funding. Both funders and implementers tend to prioritise delivery of actions and it is difficult to allocate sufficient resources to a robust outcome monitoring approach. Of course it is clear to all concerned that this may be a false economy given the need for adaptive management and the role of evidence in both improving and re-targeting management interventions and providing justification for further funding. Sustained community interest and commitment is also dependent on an understanding of performance over time and demonstration of the returns to individual, community and partnership efforts.

Notwithstanding the relative lack of quantifiable outcome measures to date there is considerable anecdotal evidence that the project commenced delivery of multiple benefits in a cost-effective manner. Through the use of local knowledge and volunteering many of the actions are being delivered at low cost, and many specific and continuing achievements of the action programmes can be listed. The increased social capital and trust between project partners also lowers the transaction costs of coordinated and collaborative actions, helping the champions to maximise their effectiveness on the ground.

BFH and BREA have used the RIPPLE project as a vehicle to engender shared responsibility for the management of the river across the community as a whole, helping solutions to be developed with less conflict and animosity. For example, both household and agricultural sources of water pollution have been identified and targeted with remedial actions, actively avoiding the allocation of blame ('finger pointing') for any one sector. This process has helped to heal long-standing conflicts and has prevented stakeholders from adopting defensive positions; a standard reaction in a situation characterised by a culture of blame. Conflict also appears to have been avoided through the gradual process of trust-building between stakeholders, facilitated by the provision of technical information where necessary. BFH has acted as a catalyst for knowledge transfer throughout the process, acquiring and disseminating scientific information to inform debate. Individuals have also been assisted to access information for themselves. While BFH and BREA are not scientific research institutes, or academic bodies, their long-standing application of applied science within the Ballinderry system and a reputation for delivery of high-quality work has gained them scientific credibility within the local community.

What is also significant in relation to the building of trust is that BREA is perceived to be a neutral information broker. This is a position BREA strongly protects, actively maintaining what the RIPPLE project coordinator describes as a 'safe distance' between BFH, BREA and any given statutory authority. BREA's neutrality has been enhanced by the transparency with which the RIPPLE project has been delivered, with all plans and processes being fully open to public scrutiny. It is also worth noting that as a charitable trust BREA is responsible for ensuring that all the projects it undertakes are solely for public benefit and open to public scrutiny.[6] Charitable status helps BREA to build trust with the local community, positioning BREA as a provider of public goods with no vested private interests or hidden agenda.

Explicit within the community development approach adopted by the RIPPLE project has been a desire to embed skills, capacity and a network of resources within the community capable of ensuring the long-term sustainability of the work programme. There have been some notable successes. For example, at the instigation of one of the RIPPLE champions the project has led to the creation of a new community volunteer group in the river system. The CURE Group (Clean-Up the River Environment) has been established to carry out conservation and access work on the river. In order to provide the group with development support and insurance provision, CURE has been enrolled as a member group in BREA. This will provide longevity to the group as well as contribute to the ongoing work of BREA in the catchment.

Despite the growing empowerment of the local community the dependence on project and grant-based funding meant that the sustainability of the programme beyond existing WWF and HLF core funding was uncertain. BREA continues its traditional river restoration work but may be unable to finance its RIPPLE project coordination role, which as noted above has been vital to the mobilisation and coordination of community activity. The project coordinator has been trained as a facilitator and convenor of community development activities but at time of writing had not yet had time to train the RIPPLE champions in these skills. Even given training it is questionable whether part-time volunteers operating from home without financial and administrative back-up can lead and coordinate catchment management planning and implementation on this scale.

Given these uncertainties concerning finance and capacity one could question whether the RIPPLE project should be fully subsumed into the official river basin district planning process being led by the NIEA. This would provide the project with statutory legitimacy and could unlock funds to continue coordination of delivery on the ground. Yet, who should then undertake the coordination role? Should it be the NIEA or should the stewardship by BREA continue? One can anticipate that if the NIEA took control processes of decision-making would change, with the risk that community empowerment, ownership and trust might be weakened. If BREA remains in the coordinating role could this be squared with NIEA priorities, standard mechanisms for the accountability of public expenditure and BREA's neutrality in the eyes of the local community? It is almost certain within the context of current statutory arrangements that a NIEA-led process would place less emphasis on the public access and education priorities set out in the RIPPLE action plan because these are lesser priorities under the Water Framework Directive. Yet it is access and education that have been the leading drivers for change for a large proportion of the local community. The challenge is to evolve governance arrangements that allow initiatives such as the RIPPLE project to be incorporated within the WFD planning process without losing community involvement, the intermediary and brokering functions provided by such as BREA, and without weakening communication, trust and collaboration between all parties.

It follows that future NIEA catchment planning should seek to avoid setting boundaries on the scope of topics to be included. Of course, publicly funded catchment management plans will have to address science-led priorities pertinent to the requirements of the WFD, but these plans should not preclude other community development initiatives and priorities. Finance for such community-derived activities might be managed independently from the public sector, for example within catchment trust funds, and be derived from multiple sources. For example, BREA could coordinate and manage these community-led activities and would work closely with the NIEA where overlap existed between community-derived objectives and WFD priorities. However, it would be necessary in the interests of financial sustainability and capacity-building for sufficient core funding to come from the public purse for BREA to perform this role.

## Acknowledgements

Thanks are due to Claire Cockerill, Ann Marie McStocker and Laurence Smith for helpful comments and editing of this chapter.

## Notes

1 Maps and more description can be viewed at the BREA website (http://www.ballinderryriver.org/pages/wheretofish.html).
2 The EU Water Framework Directive requires member states to identify river basins (or catchments) within their territory and to assign these to River Basin Districts (RBDs), which will serve as the 'administrative areas' for coordinated water management. A cross-border basin covering the territory of more than one member state must be assigned to an 'international RBD'.
3 The Budapest Declaration, 2004, defines community development as 'a way of strengthening civil society by prioritising the actions of communities and their perspectives in the development of social, economic and environmental policy'.
4 Appreciative inquiry is a form of action research that originated in the United States in the mid-1980s and involves an assessment of what enables human systems to be most capable in environmental, economic, societal, political and technological terms (Rural Community Network, 2008).
5 The Catchment Management Forum is formed from members of the local community and should not be confused with the project advisory group referred to earlier, which has an advisory role but no input into the vision and direction of the project.
6 At the time of writing all 'charities' in Northern Ireland are held to account by HM Revenue and Customs but operate under largely the same principles as registered charities in England and Wales, regulated by the Charity Commission. The Charity Commission will be established in Northern Ireland as a function of devolution.

# References

BREA (2010) *An Introduction to the Ballinderry River*, Ballinderry River Enhancement Association, Cookstown, Northern Ireland.

Budapest Declaration (2004) *The Budapest Declaration: Building European Civil Society through Community Development*, Declaration of the international conference 'Building civil society in Europe through community development', 25–28 March 2004, Budapest, www.communitydevelopmentalliancescotland.org/documents/inPractice/Budapest Declaration.pdf, accessed 22 August 2013.

Fell, D., Austin, A., Kivinen, E. and Wilkins, C. (2009) *The Diffusion of Environmental Behaviours; The Role of Influential Individuals in Social Networks, Report 1: Key Findings*, A report to the Department for Environment, Food and Rural Affairs, Brook Lyndhurst, Defra, London.

Life Long Learning UK (2009) *National Occupational Standards for Community Development*, Life Long Learning UK, London.

Rural Community Network (2008) *Appreciative Enquiry*, Fact Sheet 12, Rural Community Network, www.ruralcommunitynetwork.org/publications/publicationdocument.aspx?doc=102, accessed 22 August 2013.

WWF (2008) *RIPPLE: Visions for the Ballinderry*, www.assets.wwf.org.uk/downloads/vision_booklet.pdf, accessed 22 August 2013.

WWF (2009) *RIPPLE: A River Action Plan for the Ballinderry*, www.assets.wwf.org.uk/downloads/wwf_ripple_brochure_final_layout_1.pdf , accessed 22 August 2013.

# 9 Opening up catchment science

## An experiment in Loweswater, Cumbria, England

*Claire Waterton, Stephen C. Maberly, Lisa Norton, Judith Tsouvalis, Nigel Watson and Ian J. Winfield*

## Introduction

This chapter reports on an attempt by academic researchers, local residents, businesses and institutional stakeholders to think through and carry out catchment science, catchment management and catchment participation simultaneously. 'Understanding and Acting in Loweswater: A Community Approach to Catchment Management' was a project supported by a three-year Rural Economy and Land Use (RELU[1]) grant and took place between 2007 and 2010. It involved the creation of a new body of lay and scientific research about the catchment of Loweswater, Cumbria, within the Lake District National Park in north-west England. It also supported the creation of a new 'social mechanism', the Loweswater Care Project (LCP), which drew in, supported, scrutinised, criticised and monitored this research. All of the research was connected, sometimes directly, sometimes indirectly, to a persistent problem – the presence of potentially toxic blue-green algae (cyanobacteria) in Loweswater lake (the village, the catchment area and its lake share the same name). We describe below the origins and formation of the LCP, the ideas and commitments that underpinned it, and the consequences of its work to date.

## Context and drivers

The 2007–2010 project was led and carried out by an interdisciplinary team of natural and social scientists from Lancaster University and the Centre for Ecology and Hydrology (CEH), Lancaster, one 'community researcher' (a farmer based at Loweswater), local residents, farmers, and institutions with responsibilities for environmental quality regulation and policy. The research aimed to improve the way that these relevant actors understood and acted upon the occurrence of potentially toxic blooms of blue-green algae in the lake. The researchers also wanted to experiment in the development of more inclusive and integrated forms of catchment management, as called for by the European Union (EU) Water Framework Directive (WFD). To do this they drew on ideas of public participation in science and policy-making that long preceded the EU Directive (Tsouvalis and Waterton, 2012). Water quality had been deteriorating in Loweswater over several decades up to and into the 2000s, and conventional means of tackling the problem (scientific monitoring and regulatory interventions from 2001 onwards) appeared to have had little impact by the time the project began in 2007. Blue-green algal blooms, which are normally unusual in the colder months, were by this time becoming a regular occurrence during the winter. The problem seemed to be intractable and prompted both concern and debate within the local community and the institutions involved in the management of the catchment.

The research addressed blue-green algae in Loweswater through both 'inter-disciplinary' and 'participatory' methodologies. It aimed, first, to create a mechanism that would enable decision-making by local residents, institutional stakeholders and social and natural scientists together. Decision-making had as its objective a deliberately broad goal – the long-term ecological, economic and social sustainability of the Loweswater catchment. Thus, the LCP became a forum that opened up, rather than narrowed down, questions about what is at stake, ecologically, economically and socially for Loweswater, and possibly for other places like it. Second, the research aimed to *carry out high-quality interdisciplinary research* in order to produce a catchment knowledge base to inform such decision-making. The latter included research into upland farm economies, land and water ecology, institutional 'governance' and responsi-bilities for land and water quality, local understandings and knowledge of Loweswater, and socio-economic and cultural challenges faced by the residents. Findings from this research were shared via the LCP, the social mechanism created for the project. The LCP consistently drew local residents, institutional stakeholders and researchers together for challenging debates over a two-and-a-half-year period. Although the RELU project itself has ended, the LCP is now sustained and directed by local residents.

## Catchment background

Loweswater is situated in a relatively quiet area of the Lake District (in terms of visitor numbers) and the catchment was previously designated as a 'quiet valley' by the Lake District National Park Authority (LDNPA).[2] Relative to other Lake District lakes, Loweswater is one of the smaller, shallower lakes and has been characterised as 'eutrophic' (Maberly *et al.*, 2006, 2011). The catchment covers a land area of 7.6 square kilometres which feeds into the lake, itself 0.6 square kilometres in area. The lake is owned by the National Trust (NT) while the land area draining into the lake directly and via a number of streams comprises a mixed lowland/upland partially wooded catchment with steep-sided valleys to the north-east and south-west and gently sloping or level fields at either end of the lake.

Uniquely for a lake in the English Lake District, Loweswater drains towards the centre of the Lake District and into another lake (Crummock Water) at its south-eastern end. Loweswater has a long residence time for a Lake District lake of its size, and water entering the lake will remain in it for an average of

150 to 200 days. It is rich in submerged macrophytes (aquatic plants) while the shores include only small patches of species-rich emergent vegetation. It is used infrequently for recreational fishing, mainly for brown trout (*Salmo trutta*), and it produces only low catch rates, although in the past the lake has served as an important recreational fishery. The main inflow enters Loweswater at the north-western end of the lake after passing through lowland farming and sparsely populated residential areas. Currently there are eight farm holdings that have 'in-bye' pastures (high quality fertilised grassland) inside the catchment. The catchment has a population of approximately 45 permanent residents, and hosts around half that number again of visitors on a year-round basis in different forms of lodging around the lake (a camping barn, bed and breakfast accommodation, a small hotel and rental cottages).

Loweswater is afforded no special designation apart from its inclusion in the Lake District National Park. As part of the former Lake District Environmentally Sensitive Area (ESA) all farmland was at one time included in the ESA agri-environment scheme enabling farmers to access funding for management of farmland including capital works such as hedge re-creation. Most ESA agreements finished around 2008, if they were not renewed at that time. Renewed agreements finished during 2013. Currently, upland areas in the catchment fall under the moorland designation made by the Rural Payments Agency in England. This designation influences payments made under the EU Single Farm Payment Scheme and under new European agri-environment schemes brought in to replace the ESA scheme. Lowland agricultural areas, known as 'in-bye' land, are classified as 'Severely Disadvantaged' because of their low agricultural potential.

### Water quality and quantity problems: the issues and their severity

Loweswater experiences regular blooms of blue-green algae some of which can be toxic to both animals and humans under certain conditions (Codd, 2000; Maberly *et al.*, 2006). In the past, the NT placed warning signs around the lake advising people to keep themselves and their dogs away from the water. Algal blooms are a major water quality issue for Loweswater, affecting the use of this amenity by visitors, local residents, livestock and other animals. Water quality in the lake has been estimated to be 'Moderate Status' under the EU WFD classification (Maberly *et al.*, 2011) and long-term lake monitoring data show that the blooms are a response to high phosphorus (P) levels in the lake (Maberly *et al.*, 2011). As it is a small rural catchment the primary sources of nutrients to the lake include septic tanks serving residential and visitor accommodation, livestock farming and fertiliser application.

### Existing organisational and institutional structures concerning water quality issues at Loweswater

Despite being a relatively small catchment area, the governance and management arrangements for Loweswater are complex. The farmers make key land management decisions, but there are several public and charitable organisations which can also affect how land and water are used and managed by setting the policy context, establishing regulations and offering incentives. The Environment Agency (EA),[3] Natural England (NE)[4] and the LDNPA are important players because of their statutory responsibilities and powers. In addition, as the owner of the lake itself and a proportion of the surrounding land area, the NT, as outlined above, is a significant institutional 'actor' with a particular interest in maintaining and improving water quality at Loweswater.

The EA has wide-ranging statutory responsibilities for the protection of the environment, including water resources, and the promotion of sustainable development in England and Wales. Among the legislative controls it has available for the control of diffuse nutrient inputs from farming are, for example, the Water Resources Act 1991, the Environmental Permitting Regulations 2007, the Nitrate Pollution Prevention Regulations 2008 and the Control of Pollution (Silage, Slurry and Agricultural Fuel Oil) Regulations. However, following recognition that farming is a key source of diffuse pollution the EA has called for a new approach to the management of water quality using a 'whole catchment approach'. This places strong emphasis on the greater use of voluntary measures which can provide 'win–win' solutions for farming and the water environment, and arguably lends itself to a decision-making mechanism such as the LCP.

At a European policy level, the EA implements the EU WFD across England and Wales. The WFD requires all inland and coastal water bodies to be at or reach defined standards for 'good ecological status' by 2015, with subsequent six-year cycles of river basin planning and management to improve conditions where initial ecological and chemical targets are not achieved. For the WFD, Loweswater is included in the Derwent catchment area, which is part of the North West River Basin District (NWRBD). In the NWRBD, 70 per cent of surface waters (512 separate water bodies) were classified as failing to meet good ecological status in 2009. However, planning focuses on large geographical areas and the amalgamation of water bodies for assessment purposes has meant that particular conditions and water quality problems in small lakes such as Loweswater are obscured and effectively 'lost' in the process. Even though Loweswater is currently at only 'moderate ecological status' it seems unlikely that the first WFD river basin planning and management cycle which runs until 2015 would have a significant impact on the Loweswater catchment.

The England Catchment Sensitive Farming Delivery Initiative (ECSFDI), which nationally is a significant funding and advice mechanism for addressing diffuse pollution and water quality, is included in the responsibilities of Natural England (NE). The ECSFDI involves close cooperation between farmers and NE in the development and implementation of soil and nutrient management plans and effective applications of manures, both aimed at reducing inorganic fertiliser inputs. In addition to advisory services for farmers, a capital grants scheme is available in designated priority catchments. However the ECSFDI has only been applied to selected 'priority' catchments in England. Following discussions in 2008–2009 with representatives from NE regarding the possibility of Loweswater being included in the priority area, the LCP was informed that this would not be possible. This, paradoxically, was due to the fact that Loweswater farmers were already working together on the problem of diffuse pollution. The ECSFDI was targeted at those areas where awareness and action on diffuse pollution was limited or non-existent, whereas Loweswater was a site where there was already significant concern and desire to act. The exclusion from the ECSFDI at this time was seen as unfortunate by Loweswater farmers who were keen to attract capital grants to improve the financial feasibility of on-farm improvements to manage run-off, slurry and other known sources of phosphorus to the lake.

Other programmes of direct relevance to water quality improvements from agriculture include the Single Farm Payment of the European Common Agricultural Policy and associated cross-compliance requirements with environmental standards, and the Rural Development Programme for England which includes the Uplands Entry Level Scheme (UELS) and the Environmental Stewardship Entry and Higher Level Schemes (ELS and HLS). ELS and HLS replace the Environmentally Sensitive Areas scheme (ESA) under which farmers in Loweswater have previously received payments for management which lowers environmental impact. The ELS is a broad and shallow scheme, not tailored to the improvement of high quality areas, whereas HLS allows a more flexible and tailored approach for improvement of high-quality areas. The potential for HLS to play a part in improving water quality at Loweswater is referred to later in this chapter.

The LDNPA's statutory responsibility is to conserve and enhance natural beauty, wildlife and cultural heritage, and to promote understanding and enjoyment of the park area by the public while also fostering the economic and social well-being of local communities. As such, the condition of Loweswater and the impacts of farming and other land-based activities on water quality are of direct concern to the authority and its partners. In contrast to previous plans, the management plan for the Lake District National Park 2010–2015 was produced by 23 partnership organisations, which included borough, district and county councils plus organisations such as the EA, National Farmers Union (NFU), the NT and NE. It relates to the park area and not just the authority itself. The Loweswater Care Project (LCP) is specifically named in the plan as one of 14 lake and valley catchment initiatives meant to guide and influence the management of the landscape over the five-year period. Prior to 2010, there was no explicit acknowledgement in the management plan of the water quality issues experienced at Loweswater.

Strategic Activity Number 22 in the 2010–2015 Plan – to improve the quality of surface waters in the National Park – includes aims particularly relevant to the management of land and water at Loweswater. It is to be achieved by undertaking a comprehensive lakes-wide programme of surface water quality improvements, led by the EA. One-year action plans and five-year business plans are agreed to tackle water quality issues. However, the plan states that these 'will initially be for the priority catchments of Bassenthwaite Lake and Windermere but will seek to cover the whole Lake District in the future'. Thus at the time of the research project it was recognised that it might be a number of years before Loweswater will benefit from this main water quality improvement programme in the Lake District.

The NT takes an active role in promoting a more integrated approach to the management of water, land and related natural resources (National Trust, 2008). Key recommendations include the management of pollution at its source rather than traditional 'end-of-pipe' treatments, which are expensive and energy intensive. It also argues that 'it is time to move away from fragmented land and water management to embrace a new approach that respects natural river catchments and their processes, and considers our impacts upon water along its entire path from source to sea' (National Trust, 2008, p. 24). To achieve this, the NT has called on other public and private interests to adopt the same principles. However, targeted action to this effect has not yet been applied to Loweswater.

We note for the record here that both the LDNPA and the NT have policies which could be used to underpin a small amount of investment in Loweswater for potentially very promising environmental quality returns, as well as publicity concerning timely and well-targeted 'action'. Much of the research needed to underpin such actions has already been done through CEH and the RELU project; monitoring could be continued through the LCP at very little cost; and simple, inexpensive infrastructural improvements could make a significant difference to the amount of phosphorus reaching lake waters.

## Getting started

### How the Loweswater Care Project started: key events, organisations and people

The research and participatory mechanism described below built on a prior project initiated by a group of approximately ten farmers aimed at tackling the algal bloom problems. It was instigated and led by the late Danny Leck, a local farmer, and was called the 'Loweswater Improvement Group' (2002–2003). Through part-time work with an organisation aiming to help rural businesses after the 2001 foot and mouth livestock disease crisis (Rural Futures in Penrith), Danny Leck became aware of potential funding sources that farmers could access to help them address nutrient losses from their farms.

The Loweswater Improvement Group did not arise purely out of concern for the lake, although that was a key motivation. Farmers also wanted to pre-empt interventions by the EA which had recognised that Loweswater was unlikely to reach the water quality standards required by the EU WFD. As a result, the EA had begun to serve notices on a number of properties instructing them to check and address any problems with their septic tanks. Farmers were aware that their farming practices might also come under scrutiny from the EA and this motivated them to take action.

Another key organisation driving change in the catchment was the NT. Concerns about lake quality in the early 2000s began to sour its relations with local farmers. With the NT keen to ensure that improvements were made, and at the Loweswater Improvement Group's request, the NT provided funding for soil samples to be taken in 2003 and the results were used to advise farmers on fertiliser application.

In 2003/4 the LDNPA, in drawing up a management plan for the park, held consultations with a range of stakeholders. Two Loweswater farmers (Danny Leck and Ken Bell) attended one of the meetings and spoke about their actions at Loweswater. This gained the interest of a scientist (Lisa Norton) from CEH and eventually led to work in 2005 to investigate the impact of farming practices on Loweswater funded by NE and the NT (Maberly et al., 2006). Through the period 2003–2007 farmers and locals worked together, sometimes with scientists, to gain funding for practical actions to address the lake's pollution. These included the installation of new septic tanks at several properties at the north end of the lake, the installation of new slurry facilities and a small reed-bed, and new systems for the separation of rainwater and slurry in farmyards.

In 2004 a scoping study[5] explored the possibility of expanding the farmer-based Loweswater Improvement Group to include other relevant stakeholders and local residents (Waterton et al., 2006), and in 2007 this expansion formally took place through the initiation of the 2007–2010 project.

## Underpinning rationale and leadership

One of the three principal aims of the 2007–2010 project was to create a new 'social mechanism' which would comprise researchers, the local community and other relevant stakeholders. The 'Loweswater Knowledge Collective', as it was initially called, was envisaged as a way of sharing expertise, collective learning and working together to identify solutions. It would adopt a holistic and catchment-based perspective, and thus would incorporate many of the aims and characteristics of integrated catchment management (ICM). It would also seek to recognise and exploit the benefits of public participation in decision-making, while being informed by recent social scientific critique of such processes. More specifically, participatory catchment management has often been criticised in the past for being too agency-centred and expert-led, with other important and legitimate voices often not being given adequate recognition or opportunity to have a meaningful input to actual decisions. In effect, public 'consultation' rather than direct participation in decision-making has tended to be the norm, although many such initiatives have been presented as 'participatory' by their advocates. As such, the LCP was to be set up by the local community, stakeholders and researchers *together* as part of the research project, and structured in such a way that it facilitated: the co-production of knowledge by scientific, institutional and lay persons alike; the opening up of multiple perspectives on the 'problem' at hand; and the creation of a social space in which disagreements, the struggle to create new problem definitions, and agonistic debate could take place (Tsouvalis and Waterton, 2012). The Loweswater Knowledge Collective was renamed (in June 2008) by the participants as the Loweswater Care Project (LCP).

From 2007 onwards, the LCP consisted of a heterogeneous group of people. Members were not pre-selected, making this a truly open forum. Between 2008 and 2010, the LCP met 15 times, roughly every two months, for meetings lasting up to three and a half hours. It typically attracted between 25 and 35 participants, including three to six natural/social scientists from Lancaster University/CEH Lancaster, two to five agency representatives from NE, the NT, the LDNPA, the EA, and local residents and farmers among others. The agenda for each meeting was driven by LCP participants, and there was not a single strong 'leader' of the group. Rather, the group worked collectively, generating ideas and future proposals from within. Meetings were initially chaired by Lancaster University/CEH researchers or Ken Bell (a local farmer employed one day per week on the research project as a 'community researcher'). Since January 2011 meetings have been organised and chaired by residents living within the catchment and a LCP Steering Group consisting of eight people has been established, the majority of whom live in the catchment or locally. Under this volunteer steering group, the LCP decided to call itself the Loweswater Care *Programme* (rather than 'Project'), and to become part of the West Cumbria Rivers Trust (WCRT, 2013).

### Development of a common vision and setting of goals

Many of those involved in the LCP wanted to avoid problems that had beset other examples of public participation in environmental (and other) decision-making in the past. One such problem is the creation of a 'common vision' at the expense of allowing disagreement and heterogeneity to thrive in a group of what are, after all, people with very different perspectives. The LCP therefore had to try to balance the sense of reassurance that a bottom-up group derives from having a 'common vision' with a sense that dissent, disagreement and thinking differently, or 'outside of the box', are important and valued.

A common vision was created through the agreement of a 'mission statement' in February 2009, which read:

> The Loweswater Care Project (LCP) is a grassroots organisation made up of local residents, businesses, farmers, ecologists, sociologists, agronomists, environmental agencies and other interested parties. We work collectively to identify and address catchment-level problems in an inclusive and open manner. The LCP's vision is to gain a better understanding of the diverse challenges faced by the Loweswater catchment and together to seek economically, socially and ecologically viable ways forward and put them into practice.

This mission statement is intentionally not prescriptive about specific goals that should be achieved and during the process of creating it a lot of emphasis was placed on the need to improve understanding and to allow for dissent, disagreement and the articulation of alternative perspectives. This was something repeatedly emphasised throughout the programme of meetings and upheld by careful and deliberately inclusive chairing and facilitation.

## Approach and tools

### Processes of partnership building and co-production of knowledge

The sense of *need* for a partnership was important in shaping the LCP's approach because existing monitoring efforts, organisational and institutional arrangements, and the threat of possible EA penalties were making no palpable difference to the blooming of algae on Loweswater. The new partnership thus *had* to provide an alternative approach to those conventionally adopted in the UK to deal with issues of diffuse pollution. This meant exploring different understandings of the problem, witnessing and incorporating into the research different forms of expertise that might be relevant to the problem, and finding new ways of working together. From 2008 onwards, the LCP started to experiment by thinking critically about the *co-production* of knowledge and action. All forms of existing knowledge and expertise were considered valid, and all were critically questioned, while attempts were still made to bring together new forms of knowledge, data, understanding and experience for scrutiny and possible use. Working in this way meant that LCP participants learnt how to appraise critically, in public, many different kinds of knowledge about Loweswater and its social, economic and environmental connections (including affective experiences and memories) (Tsouvalis and Waterton, 2012). At each meeting many different forms of new knowledge were brought into the forum of the LCP to be openly questioned, critiqued and 'de-constructed' by those present. The process of partnership-building became one of questioning, debate and enquiry that, it was hoped, would lead to collective learning and possible agreement about actions and directions to be taken in the future.

Important to the LCP's way of supporting people and partner organisations to work together was the way in which the group approached 'fact-making' about algae, the lake, the catchment or other relevant 'objects'. The LCP held that, in LCP meetings:

- understandings of nature are not self-evident;
- all knowledge and expertise needs to be debated;
- uncertainties in knowledge need highlighting and accepting;
- new connections are valuable;
- doubt and questioning needs to be extended to all the LCP's representations, including scientific representations.[6]

Algae, for example, did not simply feature in the LCP as facts represented by science. Even though the LCP had a professional aquatic ecology team monitoring and modelling the state of blue-green algae in the lake, the LCP debated, agreed and also disagreed on the issues raised by the 'algae problem'. Scientific data were questioned and debated by all LCP participants. Discussions within the LCP brought more questions and connections to the fore: was the management and maintenance of feeder stream channels, or the maintenance of lake-side vegetation, connected to the algal blooms? Did algal blooms impact on fish stocks? Did they deter tourists from coming to Loweswater? Were septic tanks well-functioning or were they relatively neglected, adding to the phosphorus loading of the catchment? Such questions illustrate how the avoidance of framing the algae problem in strictly scientific terms widened the scope within which it could be addressed. It invited a more holistic approach; the water quality problem becoming set within the 'problem' of integrated catchment management.

Becoming interested in many new questions, participants also began to collect samples (the work of participant Andrew Shaw provides an example), to question prior sampling techniques (for example, questions posed by a LCP participant at the LCP meeting on 28 September 2010 regarding modelling of Loweswater), to monitor different pathways of nutrients (the study carried out by a local resident Leslie Webb), and to undertake their own small-scale research projects (see next section) to help understand the problem (Shaw, 2009; Webb, 2010). In the period 2007–2010, most of the collective partnership working that was carried out in the LCP related to 'finding out more' about the problem of blue-green algae and the way it interrelated to human and non-human ecologies and systems.

### The LCP-initiated studies

An innovative aspect of the project was based on recognition of the need for the LCP to have some autonomy. Thus a sum of money was included in the research project award that became available to the LCP during the project to commission its own small-scale research studies. This otherwise unspecified budget of £35,000 was a unique undertaking within a research project of this type. It was considered by the funders to be an important innovation in that it encouraged and empowered lay people to get involved in research and enabled them to contribute to finding out more about the issues that concerned them. It was found to have very positive impacts for the LCP and for the research outcomes as a whole. This innovation also sent a very clear signal that the LCP was a 'levelling' mechanism that enabled people from a wide variety of backgrounds and with different forms of expertise to become directly involved in scientific work.

Five studies were funded in this way, two of which were undertaken by local people: a survey of the functioning and use of all septic tanks in the catchment (Webb, 2010); and a study of attitudes to tourism and economic development in the valley (Davies and Clark, 2010). The remaining three studies included: a limnologist (lake scientist) working with a historian to compare a lake sediment sample (physical diatom data) with historical data for land-use change in the catchment (Winchester and Bennion, 2010); an aquatic ecologist working with farmers in the catchment to collect more data to understand how agricultural phosphorus indices related to phosphorus flows in the catchment; and a hydro-geomorphologist carrying out a study of the macro-scale hydrological movements in the catchment (Haycock, 2010).

## The importance of trust between partners and how it was developed

In 2007, good relations between the local actors at Loweswater were not widely enjoyed. The owners of the lake (the NT) and the farmers mistrusted each other, while relations between non-farming local residents and the NT and those between Lancaster University/CEH researchers, local residents and farmers were more variable. Over the period 2007–2010 trust between these different parties grew. In particular, the improvement of relations between local farmers and the NT was hailed as one of the main achievements of the LCP.

In part we attribute this change to the approach that the LCP took to knowledge-making and expertise, an approach that was open and simultaneously critical and reflexive about all forms of knowledge and expertise employed. Together with the lack of hierarchy designated to individuals, organisations and different sources of knowledge this engendered an enquiring and trusting atmosphere within the LCP.

Another contributory factor to the growing sense of trust between participants observed over this period was the procedures followed in LCP meetings. Meetings typically spanned an entire evening (5.30 to 9.00 p.m.) beginning with tea and biscuits and time to chat and catch up. Mid-way through the meeting all participants shared a cold buffet together. Again this provided time for informal conversation, making connections, discussion of issues and organisation of additional activities or meetings as well as bridge-building among people and organisations that had been in dispute in the past or had little prior contact with each other. LCP meetings consistently attracted good numbers and this made for a congenial atmosphere.

A further factor engendering trust may have been the attention paid to communication. All LCP meetings were advertised through the parish newsletter distributed to all residents in the parish. Invitation cards for every forthcoming meeting were sent to everyone that had previously attended or expressed an interest in the LCP, as well as to all households in the parish. Write-ups of past meetings also often featured in the parish newsletter. Minutes of the meetings were posted on the 'community noticeboard' of the research project website.

The continuity and regularity of meetings, adequate communication about meetings, a sense of purpose in bringing new knowledge and new ways of thinking about Loweswater into the forum, and the way the meetings proceeded in practice, all seemed to help to foster trust between participants.

## Processes of stakeholder engagement

As mentioned above, the LCP was made up, in part, by institutional representatives, both regular (EA, NE, LDNPA and NT) and occasional attendees (private and public bodies). In order to capitalise on this, and to make explicit the roles and responsibilities of participating institutions, one LCP meeting – proposed by local people taking part in the LCP – was organised around the theme of 'getting to know your institutions' (15 July 2009). This involved informal talks by representatives of the key agencies followed by open discussion and a question-and-answer session. This meeting indicated that all of the institutions recognised the importance of integrating land and water management within the catchment area and had begun changing the ways in which they developed and implemented policy in order to reflect this new and more integrated approach. Furthermore, several institutions were already working together to address land and water problems in parts of the Lake District. A sense of enthusiasm and commitment towards working in partnership with local communities was obvious, and this was reflected in the good attendance by institutional representatives at LCP meetings over time.

Nevertheless, it also became clear that obstacles remained that prevented institutions from putting integrated catchment management fully into practice. Each institution operates within different geographical boundaries and at different spatial scales. For example, the EA was organising its work around the North West River Basin District and the large-scale catchment areas within it. In contrast, the geographical jurisdiction of the LDNPA is divided into five 'Distinctive Areas' that reflect different social, economic and environmental characteristics. The NT manages its land and water assets (including Loweswater) on the basis of the estates which it owns, rather than in relation to catchment areas. Although the ECSFDI operated by NE takes some account of hydrologic boundaries, the priority areas identified for advice and capital grants do not correspond with the geographical boundaries used by any of the other institutions. As such, there was observed to be a fundamental institutional problem of 'spatial fit' which was hindering progress towards the implementation of integrated catchment management at a local level.

Interactions with the institutions further indicated how the complexity and uncertainty of water quality issues at Loweswater (often highlighted through LCP's critical discussions of 'the facts') presented challenges to them. Institutions tend to be more accustomed to working on well-defined problems where actions, responsibilities and intended outcomes can be quickly identified. Water quality problems at Loweswater were also perceived as not particularly serious or significant compared to conditions in other lakes and water bodies in the region (for example, the iconic Bassenthwaite Lake and Windermere). Thus, while institutional representatives were enthusiastic about the LCP and its way of working, some were unsure about how their organisation could contribute and link the insights generated through the LCP with their own organisation's decision-making processes and policy priorities. This highlights a key tension in the LCP method: while opening up knowledge-making to critical scrutiny ensures buy-in for a wide range of participants, helping to maintain a stable, public forum for debate, it may also work to de-stabilise the confidence of agencies in 'knowing what to do'.

### Awareness-raising, education and outreach activities

As a forum the LCP evolved organically from its roots in the Loweswater Improvement Project and from the inclinations, intuitions and desires of all those who were helping to form it. As such in the year 2007 it was a forum that was in a delicate state of emergence. Awareness-raising, education and outreach beyond its immediate members and stakeholders were not considered relevant at that stage, although the importance of effective communication within the Loweswater catchment itself was recognised. However, by 2010 the LCP had built up more self-understanding, more confidence and more knowledge about its main focus – the blue-green algae on the lake. Using additional end-of-project funding for knowledge exchange LCP participants in 2010 together created a booklet about how the LCP had developed and what it had achieved to date. The LCP also attracted attention from Radio Cumbria, and the BBC's *Countryfile* programme broadcast a feature about it on 25 September 2011.

So far, the LCP has not engaged in any education activities, but on several occasions members of the LCP have been invited to present to other groups (e.g. the Coniston and Crake Partnership, the International 'Living Lakes' conference held at Windermere in 2009, the Shropshire Hills Area of Outstanding Natural Beauty (AONB) Partnership, the Lake District's Still Waters Partnership, the Northern Rural Network, and the Government's Commission for Rural Communities). At the end of 2010, a workshop was held in Penrith with national, regional and local policy-makers and regulators. The aim of the workshop was to explore the LCP's experiences and achievements with these participants and to probe their attitudes to participatory and voluntary action. The experience of the LCP has also regularly featured in national knowledge exchange networks and events. The authors also engaged in further funded knowledge-exchange activities to translate the LCP's approach to knowledge-making and participation to bigger and more complex catchments.[7]

### Use of local and expert knowledge

The research project's approach – experimenting with local-level, community catchment management that sought to integrate both natural sciences (land and water) and social sciences – provided the opportunity for original methods of scientific investigation. The small studies described above represented one aspect of this in which a mix of local stakeholders and professional researchers worked together to carry out research. Other aspects of the project similarly demonstrate this collaboration. For example, the formulation and use of modelling approaches incorporated a wide range of expertise from on-site land management considerations to scientific measurement. The rationale here was that increasing local engagement with an issue can help to improve the potential for understanding the causes of the problem through provision of more accurate site-based information. Additionally, the potential for resolving the problem is increased by on-the-ground understanding of possible causes, and through engagement with those who can effect change. Thus the data sources for the modelling included considerable input from local people, right from defining the water catchment, through local measurements of rainfall, to individual management practices for septic tanks. Information about on-farm practices was a key component of the data and was supplied by farmers working with a trusted local advisor alongside the scientists. The use of an agricultural expert to interview farmers considerably enhanced the quality and depth of data obtained. As anonymity was assured, farmers were also more open about their management practices.

### Assessment of land use and farming

Established survey methods (Carey *et al.*, 2008) were employed to measure the ecological and landscape attributes of the land in the catchment by a landscape ecologist from CEH who recorded habitats, landscape features and vegetation across the catchment using a ruggedised computer. This method, while labour intensive compared to the use of, for example, earth observation data, was employed to broaden the scope of potential study beyond water quality management (although water quality is the focus of this chapter). An agricultural economic assessment of the catchment was made in collaboration with the farmers by a local farm business advisor who collected data on stocking rates, fertiliser application and other parameters needed to create a farm phosphorus budget, as well as information on farm incomes. Again, while the core use of the data was in relation to potential water quality impacts, extra data on farm incomes provided the opportunity to look at other aspects of catchment sustainability.

### Aquatic monitoring

A monthly lake monitoring programme, using standard techniques, was carried out over three years. Additionally, after considerable local consultation regarding visual impacts and other concerns, a meteorological station with lake monitoring equipment was installed on a buoy on Loweswater. Data, downloaded by telemetry, included temperature profiles, oxygen concentration, temperature, pH, conductivity at surface and depth, and surface chlorophyll *a*. Data were uploaded to the project website and made publicly available. Additional catchment data came from a previous study of inflows to the lake by CEH and from flow data provided for Loweswater and nearby catchments by the EA. Aquatic monitoring also included work on fish populations. Fish research complemented water quality monitoring and modelling exercises and formed an important part of the project because of its known interest to local and other stakeholders (Tsouvalis *et al.*, 2012; Shaw, 2009).

## Linking land management and water quality through modelling

Ecological research attempted to understand land management impacts on water quality using linked models. By modelling the catchment it was hoped to understand how nutrient loads in the lake were linked to farm management practices and to algal blooms (Norton *et al.*, 2011).

Terrestrial ecology and farm management data were incorporated into a geographical information system (GIS) enabling land management practices to be linked to the environmental quality of the land. The modelling methodology used a series of linked models to assess phosphorus run-off from the catchment to the lake and its impact on water quality. Outputs from a farm nutrient budget model (PLANET, 2013), fed into a hydrological model (*Generalized Watershed Loading Function, GWLF*; MapTech, 2013) and nutrient outputs from the hydrological model fed into the algal production model 'PROTECH' (CEH, 2013). In order to parameterise these models fully, alongside the GIS information, land management information from the farmers, meteorological data from the in-lake buoy and from residents' rainfall gauges, and hydrological data from the catchment (and where lacking from adjacent catchments) were used.

In order to test the validity of the modelling approach the models were run using current data to model nutrient inputs and potential resulting algal concentrations. Data from the automatic monitoring buoy was used to validate the results. In addition, four scenarios were explored to reflect alternative land management options and to provide information about how changes in management practices may potentially impact on nutrient inputs and algal concentrations. Non-farming scenarios included a wooded (deciduous) catchment ('woodland' scenario), and a no-input grassland scenario without livestock ('natural grassland'). Farming scenarios included 'no cattle, double sheep numbers' and 'double cattle, half sheep numbers' representing potential, though extreme, changes in the livestock composition of the catchment. Nutrient loads from septic tanks (from the LCP study by Webb, 2010) were input to the GWLF model in two ways: as a diffuse source of nutrients where phosphorus discharge from septic tanks was incorporated into the farm nutrient budget in the same way as other sources of nutrients; and as a point source of nutrients where effluent was assumed to discharge directly to the watercourse, i.e. a worst-case scenario. The models, as presented to the community, made it clear that the presence of people and livestock in the catchment comes at a cost to lake water quality, but that the cost may be minimised by improvements in nutrient management.

## Science communication

All of the scientific investigation carried out during the project was reflected back to LCP meetings. This could include quite complex issues, particularly in relation to the modelling, and scientists worked hard to make their science as transparent and understandable as possible. This was facilitated by the prior processes of engaging the LCP participants in data collection and assembly for the modelling, and in assessing the provisional results of component models (land, farming and aquatic). This all helped to make the modelling more than just an abstract exercise, and was critical, as acceptance of the legitimacy of the modelling approach was essential for its use by the LCP in informing deliberation and decision-making. The models provided information to the community, and where relevant to an individual farmer, helping to highlight the impacts of septic tank and farm management on water quality. Behavioural change in response to the information provided by the models was not formally measured but we do know that the farmer whose practices appeared to be resulting in excess nutrient loss addressed the issue and that community awareness regarding septic tank management was heightened.

## Monitoring

Much of the detailed water quality monitoring carried out by CEH depended on the research project funding from 2007 to 2010 and thus could not continue indefinitely. Monitoring that is continuing is the CEH 'Lakes Tour' that takes place seasonally every five years (Maberly *et al.*, 2011). Local residents have carried out voluntary water quality monitoring since 2011. The EA also take measures of Loweswater's 'in' and 'out' flows and this will contribute to this dataset if continued on a regular basis. Ongoing rainfall measurements in the catchment provide accurate data which can help in understanding nutrient pulses to the lake (subject to the availability of temporally consistent data on lake nutrients). In addition, a local resident is continuing phytoplankton counts to supplement the EA monthly chemistry analysis. Continued monitoring of farm nutrient inputs and losses will depend on funding for soil sampling and farmers' use of nutrient budgeting tools. While the ECSFDI catchments have provided courses for farmers to learn how to use farm nutrient budget models, this training has not yet been made available to farmers outside of those catchments. Similarly there is no funding available for farmers to have soils analysed regularly, although potential savings in fertiliser applications do provide some incentive for this. Adequate and cost-effective monitoring will be important to the future activities of the LCP but it remains to be seen how this can best develop.

# Getting things done

## *Evolution of the initial organisational and institutional structures into an operational phase*

The LCP remains a community-based and bottom-up organisation, which since January 2010 has been organised and facilitated purely from within the catchment. For example, the Loweswater farmer who was employed one day a week from 2007 to 2010 as a 'community researcher' in the research project remains involved, and became a member of the LCP Steering Group. After its formation the steering group aimed to act upon the 'knowledge base' generated during the 2007–2010 project to continue to gather evidence as a basis for decisions and action, and to explore practical ways to improve the condition of the lake. They obtained a grant from the Catchment Restoration Fund of the Department of Environment, Food and Rural Affairs (Defra) and decided to become part of the West Cumbria Rivers Trust (a charity). This ensured that further collective work would be financially supported for the following few years at least. Further funding for farm infrastructural works through the ECSFDI has enabled at least one Loweswater farmer to re-organise the storage and disposal of animal waste on the farm site.

As a minimum, a clear and valuable function for the LCP has been to provide a forum in which controversial issues can be aired, opened up, worked upon collectively and critically examined. The possibility of doing this encourages good and trusting relationships, enabling issues to be resolved promptly and efficiently. However, while many members of the community are actively working to improve water quality through the LCP, others in the catchment are not fully engaged with the process and it may be that regulatory compliance is necessary to effect changes in those cases (unless those individuals can be drawn into the LCP more effectively). The LCP is a forum in which it can be recognised that additional action may be required from the regulatory authorities also. The trust that has been built up between parties in the development of the LCP is a valuable asset, and remains vital for the future of the LCP.

# Outcomes

## *Measures of success: outcome and process indicators*

The LCP mission statement above sets out some long-term ambitions but the two most obvious indicators of its effectiveness are successful community and stakeholder engagement, and a lake with good water quality. We give some details of these below but we note first that the former may appear to have been relatively quickly achieved and quantifiable, although in practice long-term continued engagement is what is required and this is not guaranteed. The latter may also take a long time to achieve. There are some good indications in recent lake monitoring data (see Bell *et al.*, 2011) but there may be few interim indicators of progress because of changes in land management and other behaviour that are still required, and because of the potential lags in the response of the lake (e.g. to declining phosphorus inputs), as the natural system includes phosphorus 'recycling' from sediments in the bottom of the lake, among other complex ecological factors.

In terms of stakeholder engagement it is possible to gauge the level of involvement and interest through the number of organisations and individuals attending and contributing to meetings. The degree to which communication has improved between members of the group is also important. This may be measured in terms of the production of group outputs and joint decision-making. In this case outputs include a mission statement, planning documents, funding decisions for the small research projects and actions taken by the group or individuals within the group. At Loweswater a number of group actions preceded the LCP, including action by the farmers to set aside land adjacent to the lake from intense livestock production, and sourcing of co-funding to improve slurry holdings, separate waste water from rainfall on particular farms and replace septic tanks on private and commercial properties. Under the LCP itself, achievements in terms of engagement include the stakeholder meetings, and improved relations between the NT and farmers leading to joint decision-making about the management of the lake outflow and alterations to farming practices resulting from the outputs of the lake modelling exercise. The 'foundations' have been laid for continued group activity, improved individual awareness for management and maintenance of septic tanks, and regular soil testing for more optimal nutrient management on farmland. From 2012 further work and actions were supported by the West Cumbria Rivers Trust and a grant from Defra.

In terms of lake water quality, monitoring by CEH has proved invaluable in providing information on changes in lake water quality, and only continued monitoring can enable the LCP to judge whether it is achieving its aims in future. As mentioned above, the data gathered to the end of 2010 was beginning to show a slight improvement in terms of total available phosphorus within the lake compared to the early 2000s; in 2008–2010, annual concentrations of phosphorus and phytoplankton chlorophyll (a measure of abundance) were lower and the oxygen concentration at a depth of 6–8m was slightly higher (Bell *et al.*, 2011). It is known that lake recovery may be a lengthy process given the unquantifiable nutrient loading still arising from lake sediments. It will be essential to maintain and enhance the knowledge base concerning the physical processes involved, and to continue to involve stakeholders in dialogue about these issues to support their motivation.

### Socio-economic impacts

Social and economic changes within the catchment are central to the concerns of the LCP and included in its mission statement, and there are certainly expectations that the LCP can have impact through working with those agencies responsible for social and economic development. Data about socio-economic aspects of the catchment has been collected both as part of the research project and by the LCP itself through one of the small research studies that was concerned with tourism (Davies and Clark, 2010). A key socio-economic issue in the Loweswater catchment is the low level of farm incomes, such that farmers see the need to 'maximise' productivity of their land and herds, with potential impact, in turn, on land and water quality. The LCP provided a forum in which this was set against the many positive impacts that farmers have on their environment and the social structure in the catchment. While this did not result in a change to the situation to date, broader awareness of the issues among the wider community and agencies involved in the catchment has been a positive outcome.

### *Sustaining the LCP*

To date, the LCP has been very well supported by the various statutory bodies mentioned above. Participants of the LCP seem to be keen to maintain the group and have decided that it should continue under the auspices of the West Cumbria Rivers Trust. The Catchment Restoration Fund grant of more than £300,000 will ensure that the LCP can continue to set its own goals and try to meet them into the future. Further funding may also be sought involving the entry of farmers into a group agri-environment scheme agreement (HLS) when current (ESA) agreements expire. This potential was explored with the assistance of NE as part of the research. The Higher Level Scheme would help to assist continued improvements in the catchment that will impact on lake water quality while providing farmers with a better income than available under the Entry Level Scheme.

There are further questions about whether the LCP should remain an independent group or link with other initiatives, for example, the Melbreak Communities, a community programme comprising four parishes including Loweswater, which is currently host to LCP web-based information. Other options are to further link with other Rivers Trusts (for example, the Coniston and Crake Partnership and Eden Rivers Trust, and Rivers Meet in Cockermouth). Further possibilities include a whole valley farmers' group or a group consisting of farmers, business representatives and agency representatives (such as already exists in Patterdale, Cumbria).

## Concluding reflections

### *Achievements*

Three key achievements are considered to stand out from the experience of the LCP:

*1   Doing science together and using other forms of knowledge-making to lead co-enquiries and to co-research new issues (co-production of knowledge).*

The LCP was a deliberately reflexive organisation, bringing together knowledge and action, but also critically questioning the procedures and tools enabling it to learn, and the knowledge and actions being produced (Rabeharisoa and Callon, 2004). In practice, this meant that items brought to the LCP for discussion were heavily scrutinised and never taken as self-evident. Uncertainties and ignorance were often highlighted, as was the need to make decisions in the face of such conditions rather than delaying actions in the hope that things will become more certain in the future. This led to many fascinating discussions and indicated areas for further research. The five small studies described above had all been generated within the forum of the LCP. They created new collaborations between local people and university and CEH researchers, thus breaking down barriers between 'science' and 'society' in very practical ways. LCP participants were therefore all involved in one way or another in the co-production of knowledge that was considered useful in understanding the problem(s) at hand. We consider this spirit of enquiry, and the level of sophisticated questioning and critique that accompanied it in the LCP meetings, to be a key achievement of this forum.

*2   Opening up multiple perspectives on the issue or problem at hand, and refusing to allow an overly reductionist framing of 'the problem' when participants see the relevance of multiple framings or connected problem definitions.*

> Well I mean if we have a dead lake we've got a dead community because it's not just the farming you know, we've got a camping barn, hotels, tourism … There's a lot of other income comes into the valley rather than just the farming now.
>
> (Interview with Loweswater resident, 2008)

As the quote above illustrates, many things in Loweswater depend on the environmental health of the lake, and one commitment of the LCP was to encourage participants to contest any particular framing of the issue of blue-green algae on Loweswater. This opened a space for them to articulate what Wynne calls a 'societal definition' of the issues of public concern, rather than imposing a scientific definition of 'the problem' from the top down and from the outset (Wynne, 2007, p. 108). In practice, this meant that public definitions of 'the problem(s)' were encouraged and questions as to what is relevant to the issue of blue-green algae in Loweswater remained open. This led to the consideration of a wide range of connecting issues including: farm livelihoods and farming futures in the Cumbrian uplands; the changing policies of the National Park and Natural England and the way these affect places like Loweswater; household detergents and the possibility of a catchment-wide change to low phosphorus dishwasher tablets; the effects of algal blooms on tourist visits; the sensitivity of food chains and the aquatic ecology of Loweswater.

Taking a very open view of what is relevant in thinking about environmental quality and environment–society interactions in Loweswater allowed an ecological, scientific, and regulatory framing of the blue-green algae problem to connect to other framings (economic, sociological, cultural, even philosophical). This led to a cycle within the LCP meetings which moved from the algal problem to wider catchment issues and back again, an intellectual 'opening up and closing down' (Stirling, 2005, p. 218) that over time began to characterise the 'rhythm' of the LCP.

*3   Creating a forum in which contestation, disagreement and agonistic struggle to define issues and problems can legitimately take place.*
One of the most significant concerns found in critiques of stakeholder participation is the idea that participation often consists of little more than a public relations exercise designed to persuade and mollify the public, and to gain their support for previously identified 'expert' discourses, rather than let that public inspire alternative ways of thinking. In contrast a third strength of the LCP has been the way that it encourages different voices to be heard, alternative perspectives to be offered and disagreements and conflicts to be aired (see Tsouvalis and Waterton, 2012 for an account of such 'agonistic' practices). This commitment to encouraging diverse and conflicting perspectives to be articulated meant that, in the long run, it became possible for all in the LCP to consider some of the inequities and imbalances that exist in Loweswater – for example, the reality that Loweswater is a place characterised by increasingly elite consumption. It is no longer a predominantly productive landscape but one that is consumed, by Lake District tourists and wealthy retirees. There exists, in fact, very little economic opportunity in the catchment for those who are not retirees or farmers. By exploring this, LCP participants have been able to appreciate, for example, that farm households making a marginal living from sheep and beef farming have, in the past, been too readily blamed for spoiling a picture-perfect Lake District scene, through their supposedly over-zealous farming practices. Through the LCP, and particularly through the representation of local farmers in meetings and the encouragement of farmers to say their piece, participants have become more interested in understanding farming trends and practices to try to find ways of ameliorating phosphorus flows from farm holdings to the lake, without attaching blame or finger-pointing.

### Weaknesses, continuing challenges and some lessons from failures

The 'openness' of the LCP to problem-definition as described above might be perceived by some as a weakness, not least for those used to more conventional scientific approaches to catchment management. Occasionally, it led to outbursts of frustration as, for example, when a participant would suggest that the group, as a whole, had 'lost focus' on blue-green algae while pursuing these other connections. On other occasions participants would acknowledge that it was impossible to think about the algae 'in isolation'. The connections made and their relevance seemed to be increasingly appreciated over time, as a more complex and composite picture of Loweswater and its algae began to emerge through investigations and discussions played out through the LCP.

The LCP experience has also shown that doing science with the public and fostering trusting relations between scientists, farmers, institutional stakeholders and the public is a time-consuming and long-term commitment. It involves a considerable investment of time in 'finding out' what the issues of concern are, how they are connected and how they might best be addressed and by whom. Again this is not a direct weakness as such but it can slow down diagnosis and action. On the other hand, if a problem is only narrowly defined and action is imposed quickly, the solution may not find acceptance among the community or may prove misdirected. It might even exacerbate a 'problem' that is poorly understood.

Continuing challenges include defining catchment improvement actions, and to strengthen the partnerships needed for many actions. The latter raises the question as to whether the LCP can convince relevant institutions to prioritise Loweswater; and for the institutions it raises the question as to whether they will be able to adapt their often larger-scale focus and systematised programmes to the specific needs of this small catchment. A further challenge extends beyond Loweswater. Despite Loweswater's small scale and relative simplicity as a catchment, the authors think that knowledge of the way that that participation and science were effectively integrated in local catchment management through the LCP could possibly be used to improve the way the EU WFD is being implemented. Further knowledge exchange activity in 2012 explored these and other issues, producing some simple recommendations from the Loweswater project for future catchment management projects (Waterton *et al.*, 2012).

## Notes

1    The project was funded by the Rural Economy and Land Use (RELU) programme of the Economic and Social Research Council (ESRC), the Biotechnology and Biological Sciences Research Council (BBSRC) and the Natural Environment Research Council (NERC), with additional funding provided by the Scottish Government and the Department for Environment, Food and Rural Affairs.

2    A map and further information can be viewed from the website of the West Cumbria Rivers Trust (http://westcumbriariverstrust.org/areas).

3    The Environment Agency is an executive non-departmental public body responsible to the Secretary of State for Environment, Food and Rural Affairs in England and a Welsh Government sponsored body responsible to the Minister for Environment and Sustainable Development in Wales. Its principal aims are to protect and improve the environment, and to promote sustainable development, and it is empowered by law as the main regulator of discharges to air, water and land.

4    Natural England is an executive non-departmental public body responsible to the Secretary of State for Environment, Food and Rural Affairs. Its purpose is to protect and improve England's natural environment and encourage people to enjoy and get involved in their surroundings.

5    Also RELU funded.

6    These commitments to questioning were derived from the work of Latour (2004).
7    A project funded under the NERC's Water Security Knowledge Exchange Programme: 'Understanding and Acting in Loweswater: A community approach to catchment management', January–April 2012.

# References

Bell, K., Maberly, S., Norton, L., Tsouvalis, J., Waterton, C., Watson, N. and Winfield, I. (2011) *Understanding and Acting in Loweswater: A Community Approach to Catchment Management. Summary Report*, Lancaster University/CEH.

Carey, P., Chamberlain, P. M., Cooper, A., Emmett, B. A., Maskell, L. C., McCann, T., Murphy, J., Norton, L. R., Reynolds, B., Scott, W. A., Simpson, I. C., Smart, S. M. and Ullyett, J. M. (2008) *Countryside Survey: UK Results from 2007*, Countryside Survey, www.countrysidesurvey.org.uk/outputs/uk-results-2007, accessed 23 August 2013.

CEH (2013) *Lake Modelling PROTECH*, Centre for Ecology and Hydrology, www.ceh. ac.uk/sci_programmes/water/lake%20ecosystems/lakemodelling.html, accessed 23 August 2013.

Codd, G. A. (2000) 'Cyanobacterial toxins, the perception of water quality, and the prioritization of eutrophication control', *Ecological Engineering*, vol. 16, no. 1, pp. 51–60.

Davies, D. and Clarke, E. (2010) *Community and Culture: Tourism in a Quiet Valley*, Report to the Loweswater Care Project.

Haycock, N. (2010) *Hydrogeomorphological Investigation of the Main Streams Feeding Into and Out of Loweswater*, Haycock Associates, Report to the Loweswater Care Project.

Latour, B. (2004) *The Politics of Nature: or How to Bring the Sciences into Democracy*, Harvard University Press, Cambridge, MA.

Maberly, S., Norton, L., May, L., De Ville, M., Elliott, A., Thackeray, S., Groben, R. and Carse, F. (2006) *An Investigation into the Potential Impacts of Farming Practices on Loweswater*, Report to the Rural Development Service and the National Trust, Report No. LA/C02707/4.

Maberly, S. C., De Ville, M. M., Thackeray, S. J., Feuchtmayr, H., Fletcher, J. M., James, J. B., Kelly, J. L., Vincent, C. D., Winfield, I. J., Newton, A., Atkinson, D., Croft, A., Drew, H., Saag, M., Taylor, S. and Titterington, H. (2011) *A Survey of the Lakes of the English Lake District: The Lakes Tour 2010*, Report commissioned by the Environment Agency and the Lake District National Park Authority.

MapTech (2013) *Generalized Watershed Loading Function (GWLF)*, www.maptech-inc.com/ services/gwlf.php, accessed 23 August 2013.

National Trust (2008) *From Source to Sea: Working with Water*, The National Trust, Swindon.

Norton, L., Elliott, J. A., Maberly, S. C. and May, L. (2011) 'Using models to bridge the gap between land use and algal blooms: An example from the Loweswater catchment, UK', *Environmental Modelling and Software*, vol. 36, pp. 64–75.

PLANET (2013) *PLANET Nutrient Management*, www.planet4farmers.co.uk/Content. aspx?name=PLANET, accessed 23 August 2013.

Rabeharisoa, V. and Callon, M. (2004) 'Patients and scientists in French muscular dystrophy research' in Jasanoff, S. (ed.) *States of Knowledge. The Co-Production of Science and Social Order*, Routledge, London, pp. 142–160.

Shaw, A. (2009) *The Decline of the Loweswater Fish Community: Contributory Factors and Events*, Unpublished M.A. thesis, University of Lancaster.

Stirling, A. (2005) 'Opening up or closing down? – Analysis, participation and power in the social appraisal of technology' in Leach, M., Scoones, I. and Wynne, B. (eds) *Science and*

*Citizens: Globalization and the Challenge of Engagement*, Zed Books, London, pp. 218–231.

Tsouvalis, J. and Waterton, C. (2012) 'Building participation upon critique: The Loweswater Care Project, Cumbria', *Environmental Modelling and Software*, vol. 36, pp. 111–121.

Tsouvalis, J., Waterton, C. and Winfield, I. J. (2012) 'Intra-actions in Loweswater, Cumbria: New collectives, blue-green algae, and the visualization of invisible presences through sound and science', in Rose, G. and Tolia-Kelly, D. (eds) *Visuality/Materiality: Images, Objects and Practices*, Ashgate Publishing Ltd., Farnham, Surrey, pp. 109–132.

Waterton, C., Norton, L. and Morris, J. (2006) 'Understanding Loweswater: Interdisciplinary research in practice', *Journal of Agricultural Economics*, vol. 57, no. 2, pp. 277–293.

Waterton, C., Tsouvalis, J., Maberly, S. C., Norton, L., Watson, N. and Winfield, I. J. (2012) *Public Participation in Catchment Management – Experiences From Loweswater, Cumbria*, A Knowledge-Exchange Report to NERC, Lancaster University.

WCRT (2013) *The Loweswater Care Programme*, West Cumbria Rivers Trust, www.westcumbriariverstrust.org/loweswater/, accessed 23 August 2013.

Webb, L. (2010) *Survey of Local Washing Practices and Septic Tank Operation in Relation to Domestic Phosphorus Inputs to Loweswater*, Report to the Loweswater Care Project.

Winchester, A. and Bennion, H. (2010) *Linking Historical Land-Use Changes with Paleolimnological Records of Nutrient Changes in Loweswater Lake*, Report to the Loweswater Care Project.

Wynne, B. (2007) 'Public participation in science and technology: Performing and obscuring a political-conceptual category mistake', *East Asian Science Technology and Society: An International Journal*, vol. 1, no. 1, pp. 99–110.

# Part III

# Lessons for catchment and river basin management

# 10 Getting started

## Partnerships, collaboration, participation and the role of law

*Laurence Smith, Keith Porter and David Benson*

## Introduction

Chapter 2 noted the importance of the drivers that may initiate and sustain a catchment management initiative, and of the governance context in which it will operate. Interdependent with drivers and governance are the motivations and actions of the individuals and organisations that seek to protect and manage their waters. Despite their diversity, the case studies in this book illustrate these issues and help us identify key mechanisms and processes that enable successful catchment management initiatives to form and develop.

The relevant key questions identified in Chapter 2 are as follows:

**1   What are the context and drivers for catchment management?**
(Why did it start? What factors influence when and how a catchment management programme begins? What factors promote an integrated approach?)

**2   How can a catchment management programme get started?**
(How did it start? What factors influence how catchment management initiates and evolves?)

## The context and drivers for catchment management

It becomes apparent from the case studies in this book that it is too simplistic to characterise catchment management as arising either 'top-down' in response to national objectives and regulation, or 'bottom-up' in response to local concerns and grassroots activism. In reality, an interaction between both poles of this spectrum is played out through the political process and as facilitated by the prevailing jurisdictional law, multi-level and polycentric assignment of responsibilities, and funding provisions and opportunities.

In the USA, the federal Clean Water and Safe Drinking Water Acts are significant (but not exclusive) drivers of locally managed watershed programmes. Such programmes in the Upper Susquehanna Watershed (Chapter 3) highlight how priorities assigned from the highest political level for improvement of the waters of Chesapeake Bay are translated into obligations for state, city and county governments by the provisions of the Clean Water Act and oversight by the USEPA. Chapter 4 similarly shows how the public health and water supply priorities of New York City to cost-effectively secure and protect its water supply best meet the requirements of the Safe Drinking Water Act. Yet both federal Acts, with respect to catchment and water supply protection, depend for implementation and cost-effective enforcement on the acceptance by local jurisdictions of water quality standards and pollution limits, plus acceptance of the responsibility for the achievement of these, i.e. they must have local-level legitimacy. Where progress towards federal objectives is poor, further 'bottom-up' pressure may be applied by environmental activists, not least through law suits.

The accomplishments of the Hudson River Estuary Program (HREP) (Chapter 5) provide something of a contrast in New York State watershed management. Here a wide-scale and comprehensive programme covering most of a major river basin was initiated and remains coordinated by higher-level government, i.e. New York State Department of Environmental Conservation (NYSDEC), a state government agency. Despite this, the programme owes its origins and continuing success primarily to 'grassroots' environmental activism, and for its implementation depends on voluntary acceptance of responsibilities and actions at the local governmental and public level.

In Queensland, Australia, 'top-down' state government policy has evolved to encourage locally led catchment-scale natural resource management. Notably, the approach remains one of flexible support and coordination for community-governed regional and local organisations, as compared to the more directive, structured and statutory approach now prevailing in New South Wales, Victoria and South Australia. The example of the 'Healthy Waterways' (Chapter 6) illustrates how action can grow locally through iteration and evolution in response to local economic drivers and environmental concerns, but also how this development was dependent on local scientific and political leadership at a time when there were few models or templates to follow.

The examples of groundwater protection (Chapter 7) provide both similarities and differences to the examples that focus on surface water catchments considered above. As a 'top-down' regulatory driver the European Union Water Framework Directive (EU WFD) and its predecessor directives 'mirror', at least in part, the regulatory context provided by the US Clean Water and Safe Drinking Water Acts. Similarly, regional and municipal priority given to public-health concern for safe drinking water in the cases in Denmark, Germany and the Netherlands also 'mirrors' the example of the New York City Watershed. Yet locally led partnership working in each of these three cases pre-dated much of the EU regulation, and was recognised nationally and locally as the best means to meet both national legislation and local needs. In Denmark and the Netherlands, and to a slightly lesser extent in Germany, it is also apparent that recognition of integrated rural spatial planning at a catchment scale as a means to deliver multiple economic, environmental and social benefits acted as a further driver of the approaches adopted.

As a further contrast, the cases in Chapters 8 and 9 illustrate how catchment management initiatives can start primarily through 'bottom-up' processes, yet with recognition of the necessity of being both constrained and supported by the context provided by 'top-down' regulation and prevailing multi-level responsibilities. In Northern Ireland, the WWF RIPPLE project (Chapter 8) is an example of action initiated by environmental charities that won funding for their initiative from a higher level of governance. These charities were motivated to achieve both local catchment improvements and set a benchmark for good practice in river basin planning as an exemplar demonstration of how nationwide programmes should be 'rolled out' under the EU WFD. The project wished to shape both national and local agendas, but as a result became dependent on the national (and indeed EU agenda) for its financial sustainability and for at least part of its rationale and *modus operandi*.

Finally, the Loweswater case (Chapter 9) is unusual insofar as the collective research agenda of the local community and Lancaster University provided a key driver of the initiative. Similarly to the Irish case, the research agenda encompassed both investigation of how management of the catchment itself might be improved, and an exploration of how EU WFD requirements for more inclusive, participatory and integrated forms of catchment planning and management can be achieved at a local scale. Also, in both cases the projects described in this book were pre-dated and facilitated by earlier community-led action motivated by localised and 'bottom-up' activism.

### *Factors that promote an integrated approach to catchment management*

Given the logic and desirability of integrated planning for, and management of, natural resources at a river basin or catchment scale (Chapter 1), we raised in Chapter 2 the question of whether 'top-down' or 'bottom-up' origins for a catchment management programme best promote an integrated approach. That is, an approach that is comprehensive and spans boundaries of self-interest, sector, discipline and governance. The answer, as revealed by the case studies in this book, is again that it is the 'inter-play' and interaction of the 'top-down' and 'bottom-up' that is most important in this regard.

It can be observed from all of the case studies that acceptance of responsibilities for action at local level can provide the means for integration of a community's economic and social goals with water management objectives, and that delivery of both local goals and national policies inevitably requires integration of assessment, planning and action across relevant issues and sectors. It hardly needs saying that local communities that accept the mandate to meet quality standards for the raw water that passes through and leaves their vicinity will seek to minimise and compensate for any negative economic impacts and also seek to exploit any economic benefits of an enhanced environment. Development and implementation of this integrated approach locally will still require legislative direction, and administrative and technical support from scientific partners and higher levels of government. Above all, it will require the development of a shared vision and strategic goals among the multi-level actors involved.

These observations are perhaps best illustrated by the range and scope of the actions of the Delaware County Action Plan (DCAP) in the New York City Watershed (Chapter 4). Here it was recognised that farming, urban and rural highway runoff had become by far the greatest sources of water pollution in the catchment once point discharges were adequately managed. Highway ditches and drainage especially, are the primary conduits for runoff and stormwater from adjacent land and from the highways themselves. Protecting the quality of storm and other drainage water from land-based non-point sources of contamination in rural and suburban areas is a shared challenge for the community, the importance of which is generally underestimated. Thus secure protection of the water supplied by a populated catchment with a viable economy depended on the willingness and ability of farmers and other landowners, businesses, community leaders and residents to manage the sources of potential pollution that they controlled. As elsewhere, mitigation and control of non-point sources in particular depended on local management. In the DCAP, integration of water quality, economic and social objectives has been achieved through the necessary local acceptance of responsibility and the development of local capabilities, with support and guidance from scientific partners and the relevant city, state and federal agencies. However, as noted in Chapter 4, the vision that a protected water supply can best be achieved by the stakeholders who live and work in a catchment is still not universally shared by all actors and interest groups in New York City and the State.

A second, and geographically much larger example, is provided by the Hudson River Estuary Program (HREP). Its twelve 'Action Agenda Goals, 2010–2014' (Box 5.1, Chapter 5) illustrate the scope of an integrated approach to the conservation and recovery of an estuary and its catchment on this scale. This is not a 'top-down' or regulatory plan, but rather a 'mission statement' by the state and its citizens for the future of the Hudson Valley region. It is a 'bottom-up' plan insofar that it was developed through sustained consultative activities and outreach with communities along the length of the river, and depends for implementation on actions by volunteers, partnering groups and municipalities, as well as by state and federal agencies. As noted in Chapter 5, inter-municipal agreements also importantly provide the legal means for integrated regulation, planning and action across administrative boundaries. By incorporating and integrating the multiple interests of the estuary's citizens through this range of mechanisms, the Action Agenda in all its detail has become a shared vision and has consequently gained widespread acceptance.

In both of these examples from the USA, an integrated approach has been developed through an iterative and adaptive process over time. This is also well-illustrated by the Healthy Waterways programme in South East Queensland (SEQ), Australia (Chapter 6). From 1999 to 2002 the Brisbane River and Moreton Bay Wastewater Management Study (the 'SEQ Study') that had been a starting point, was broadened to incorporate rivers, inland catchments and the challenges of diffuse sediment and nutrient pollution; leading through partnership working to publication of a regional water quality management strategy. The next phase of development, 2002 to 2005, saw integration of management of the region's coastal and inland waters – including decision support tools, research programmes and collaboration arrangements – into an adaptive management cycle incorporating comprehensive data collection, analysis, planning, plan implementation, monitoring and evaluation. This planning process intensified over time so that the SEQ Healthy Waterways Strategy for 2007 to 2012 comprised twelve integrated action plans developed collaboratively by partners through consultation, research and analysis; but crucially also integrated with higher-level regional and state planning processes, leading to endorsement by the Commonwealth under its National Water Quality Management Strategy as the Water Quality Improvement Plan for Moreton Bay. The integrated approach had now evolved to encompass 'whole-of-water' cycle management, including actions to prevent water scarcity under conditions of future climate change and population growth.

Throughout this evolution it became essential for a common vision, goals and approach to be agreed. This allowed the setting of specific targets for water quality and their pursuance through a collaborative adaptive management approach. Local priorities had to accord with higher-level policy objectives so that local and regional activities became 'joined-up' and coordinated. Thus targets set locally also deliver the targets set in higher-level natural resource, land use and water supply plans, as well as being in accord with environmental protection policies and legislation. The crucial lesson here is that such an approach provides legitimacy as well as coherence, and 'opens the door' for the funding of locally determined action plans by higher levels of government.

With the help of the case study examples summarised above, this chapter has argued that acceptance of responsibility for planning and action at a local catchment scale can lead to the potential for local water quality and other environmental improvements in accordance with higher-level regulation and policy, and balanced achievement of this aim with local economic and social development goals. The case studies of groundwater protection from north-west Europe presented in Chapter 7 go a step further in illustrating that programmes and policies with a single focus such as the reduction of nitrate pollution are likely to be less effective than integrated and holistic approaches that aim to achieve multiple benefits and are capable of resolving trade-offs between interest groups and their objectives.

The case studies in this book also indicate that an integrated approach is a key element in gaining stakeholder support and additional financing to facilitate implementation. For example, they suggest that plans to protect groundwater should be integrated with those for other goals such as recreation, landscape amenity, biodiversity conservation, a growing rural economy and jobs. Solutions to diffuse water pollution based on cooperative agreements between farmers and water suppliers need to be integrated with spatial land use planning at a local or regional scale. In the cases described in Chapter 7, this type of approach delivered synergies and multiple benefits for different stakeholders, and provided opportunities to combine funding streams and achieve more effective horizontal coordination of otherwise sector-specific action plans and interventions.

Finally, the two case studies from the UK provided examples of how integrated plans can evolve at a local level. The RIPPLE action plan (Chapter 8) exemplified how local concerns and motivations can generate the impetus for a comprehensive planning process at a catchment scale. The plan was mainly implemented through community action groups, each coordinated by a champion and supported by the RIPPLE project coordinator. Fourteen champions led on subject areas or themes that ranged from the highly local and specific such as dead animal disposal and invasive weed control, to catchment-wide issues such as recreational access and engagement with business.

Chapter 9 explains how a team of academic researchers worked with a community-based catchment management initiative at Loweswater in north-west England to investigate whether and how collective learning and action would lead to the adoption of holistic and catchment-based planning, and thus incorporate at least some of the aims and characteristics of inclusive and integrated catchment management as called for by the EU WFD. One of the key characteristics of the Loweswater Care Project (LCP) that emerged was agreed practice to open up multiple perspectives on a problem at hand, explicitly allowing for multiple and holistic framings of that issue. The resulting problem-framing for their lake's water quality encompassed issues ranging from livelihoods in the Cumbrian uplands, the effects of algal blooms on tourist visits and aquatic ecology, through the policies of statutory agencies, to micro-management issues such as use of household detergents and dishwasher tablets. An ecological, scientific and regulatory framing of the blue-green algae problem therefore became connected to economic, sociological and cultural framings of causes and impacts. It also led to community recognition of barriers to implementation of a more integrated approach to solutions; not least the fact that the three leading governmental agencies of relevance were operating at different spatial scales and with differing but overlapping remits and geographical boundaries.

## How to get a catchment management programme started

The case studies in this book demonstrate that the questions of how best to initiate and manage the early stages of development of a catchment management programme are non-trivial. Collaborative catchment management can be understood as diverse organisations with relevant mandates and responsibilities working together to develop and advance a shared vision, while engaging wider stakeholders in knowledge exchange and deliberation. It should be inclusive, and thus will involve representatives of a wide cross-section of organisations, interest groups and citizens with a stake in the processes and outcomes. How does this get started?

In answering this question the case studies in this book reveal some commonalities of principles, process and approach, but diversity in organisational structures, institutions and leadership. Getting started will usually involve an open and inclusive process of partnership-building, stakeholder engagement and information-sharing. It should usually involve the development of a shared vision and goals, trust between participants, a shared and strengthening scientific knowledge base and, as far as possible, diverse and flexible mechanisms for funding.

In the USA, these processes were led, coordinated or undertaken in varying degrees and manners by key individuals in county-level Soil and Water Conservation Districts in the Upper Susquehanna (Chapter 3), county-level local government supported by Cornell University in the New York City Watershed (Chapter 4) and a state Department of Environmental Conservation (NYSDEC) in the Hudson Valley (Chapter 5). The prior existence of capacity in the form of a cadre of technical providers (see further discussion in Chapter 12), working within catchments and accepted and trusted by local stakeholders, was a pre-eminent advantage.

In Queensland, Australia, action by local scientists and city politicians evolved into a small professional secretariat supported by Brisbane City Council that similarly developed the necessary leadership and coordination role (Chapter 6). In Denmark, Germany and the Netherlands (Chapter 7), a close and effective working relationship between local government and municipal water suppliers had the capacity to initiate and coordinate development of groundwater protection programmes, each of which demonstrated the key process characteristics identified above. In Northern Ireland (Chapter 8) the key leadership and coordination roles were performed by environmental charities and a regional voluntary organisation particularly concerned to represent and promote the 'voice' and concerns of rural communities. Finally, in Loweswater in England (Chapter 9), local people themselves developed a community organisation with the help of university researchers and other agencies.

It might thus be concluded that to get started *what* is done matters more than *who* does it. However, the cases in this book also differ in scale and other characteristics. Scale matters in environmental management, particularly as catchment management processes expand. Effective catchment management requires sharing and coordination of authorities and roles between diverse farm businesses, other organisations and levels of government. Sorting out these institutional arrangements and relationships will take time at larger scales, but in all cases will be facilitated by trust, shared understanding and a broad-based acceptance of legitimacy. Thus status and standing matter in initiating and taking forward these processes. The actors who accepted the leading roles in the case studies in this book were able to demonstrate sufficient standing, independence and integrity to engage and establish trust with a wide range of stakeholders. They also held, or rapidly gained, the respect and trust of the key higher-level authorities that held statutory responsibility and power.

In the early stages of programme development, reputation and performance can overcome deficiencies in formalised legal and institutional standing but can this be relied upon? Dynamism and inspirational leadership can overcome many limitations and obstacles, but its emergence cannot be relied upon in all situations. Individuals may come and go at short notice. From the 'top' political priorities and support can waver and threaten continuation of a programme. The need for conflict resolution, for monitoring of practice and for enforcement of agreements or regulation (in cases of serial malpractice) may arise. Thus, it is crucial that however a catchment management programme gets started it soon establishes independence, integrity, trust and acceptance in the eyes of all stakeholders. It may thus require the backing of a sufficiently formalised status. It certainly becomes necessary to institutionalise both horizontal coordination at a catchment scale through communication and partnership working, and vertical coordination through assimilation of responsibilities, status and standing. An investment made in catchment management, and any early benefits gained, can be quickly squandered if a lack of standing leads to stalled implementation and a breakdown of partnership working. We consider this further below.

### Getting started and the role of law

Our introduction to this book (Chapter 1) suggested that to achieve protection of water resources at their source, local management of land and water at a catchment or sub-catchment scale should be recognised as the desirable, natural and default situation given its inherent logic and advantages. A determining aspect of this logic is the prevailing law.

The law of catchments is largely a matter of scale and function. At the whole catchment or river basin scale, water law addresses macro-legal issues. This body of law concerns major diversions of water, damming rivers, creation of reservoirs, flood control measures, and the infrastructure of water supplies and wastewater collection and disposal systems. Law at the catchment scale typically includes general rules for the control of municipal, industrial and other point sources of pollution. These aspects of water law are almost inevitably of a 'top-down' nature, but cannot encompass all water resource management challenges of concern. This book has particularly concerned the management of land, non-point sources of pollution and drainage within a catchment. The decision-makers at this level of catchment management include farmers and other landowners, businesses, local governments and even the general public. Water laws that concern this constituency include the following:

- legal tools and mechanisms for stream management including regulatory and non-regulatory measures that provide resource protection;
- local frameworks for flood risk planning and response, and legal options for reducing or mitigating flooding such as floodplain protection;
- local wetland protection and restoration;

- drainage issues relevant for property owners and municipalities including especially runoff, stormwater management and highway ditches;
- local means to protect and manage riparian buffers;
- the protection of groundwater resources through protection of overlying recharge areas and well catchments;
- the adoption and implementation of best management practices to protect water quality especially in farming;
- legal agreements through which landowners adopt management practices to protect water quality and quantity for public benefit;
- legal incentives to foster voluntary adoption of best management measures and public sanitary behaviour.

The constituencies and stakeholders concerned with such multiple aspects of the law are themselves multi-faceted. 'Top-down' management at this scale would be both impractical and undemocratic. This book has presented examples and described options to show how integrated water management can be comprehensively realised at the local level.

The organisational and institutional arrangements of an integrated and locally focused catchment management programme need to become legitimate and legally enforceable. To have this legitimate legal status the law must recognise the organisation so that it has decision-making power within its realm of responsibility, and so that its decisions can be implemented, and if necessary enforced by a higher level authority.

There are several ways by which an organisation or partnership arrangement can become legally legitimate for catchment management purposes. First, an existing entity that has legal status may take on leadership and coordination functions for catchment management. Municipalities or local governments are typical examples for this legitimacy, as illustrated in this book by Delaware County in the New York City Watershed (Chapter 4) and European cases of groundwater protection (Chapter 7). In the United States, especially, there is a deeply held aversion to the creation of 'yet another' government body. Therefore, it is essential to first seek to engage an already existing government agency. Provisions will vary by jurisdiction, but in many places such bodies have the authority to manage water resources as part of their pre-existing powers. However, powers to manage private land may be more limited, particularly in rural areas, and hence to secure land use objectives instruments such as cooperative agreements with landowners (Chapters 3, 4 and 7) and land acquisition (Chapter 4) may be employed.

Second, an already existing legal authority may enter into an agreement with another legal body to coordinate their management of land and water resources, and associated catchment interventions. This mechanism is best illustrated by inter-municipal agreements in the Hudson Valley (Chapter 5). It was also illustrated by the Coalition of Watershed Towns and New York City Watershed Memorandum of Agreement (Chapter 4), use of a Memorandum of Understanding by which nineteen county governments legally endorsed the Upper

Susquehanna Coalition (Chapter 3), and the legal formalisation of the Healthy Waterways partnership as a not-for-profit, non-governmental membership-based organisation (Chapter 6). Together these examples show that multiple mechanisms to convey legal legitimacy may be needed in parallel at different levels within multi-level governance structures, in order to achieve comprehensive vertical assimilation of legal responsibilities and status.

Third, and potentially most difficult, a 'bottom-up' initiative to protect water resources may create an organisation or partnership arrangement that must seek legal standing. Legal recognition is likely to be more difficult to obtain because the organisation does not have pre-existing legal authority or decision-making power, and will thus not already be recognised as an entity legally acting on behalf of people within a catchment, nor perhaps as a valid recipient for public funds. At the time of writing, this situation possibly remains a dilemma for the WWF RIPPLE project (Chapter 8), which posed the question of how to evolve its status to allow it to be subsumed within the national EU WFD planning process without losing its key strengths. Those strengths include involvement and prioritisation of the local community in planning, the neutral and respected intermediary and brokering functions provided by the Ballinderry River Enhancement Association (BREA), and the communication, trust and collaboration established between all parties. Other comparable challenges are faced by the LCP (Chapter 9) in wishing to gain status and sustainability through association with other government and non-governmental programmes operating at larger geographical scales, while also seeking to convince such entities to adequately prioritise Loweswater, and to adapt their wider-scale and systematised (if not standardised) action plans to the specific needs of this small catchment.

The catchment body that is created by local initiative may of course be established with legal standing to act as a facilitator or technical provider. A paradigm for such an organisation in the United States is the Upper Susquehanna Coalition (USC) in New York State (Chapter 3). Although a recognised legal entity, the USC has no regulatory authority itself and has indeed declined to accept any regulatory role despite occasional encouragement to accept delegated authority from the state. The USC is fiercely possessive of the goodwill it has with farmers, landowners and other local stakeholders including local governments. The coalition has no doubt it would sacrifice that goodwill and trust if it were to accept a regulatory role. Rather the USC provides technical support assisting landowners and local governments in particular to adopt and implement best management practices. These practices may represent regulatory requirements that are enforceable by another agency, or have a voluntary basis. Thus the USC acts as a welcomed interface between the regulators and local managers and decision-makers. In the New York City Watershed Protection Program, the Catskill Watershed Corporation and the Watershed Agricultural Council serve an analogous function. Comparable organisations are found in other countries, for example, Healthy Waterways in Queensland (Chapter 6) and the charitable-status Rivers Trusts in the United Kingdom.

A 'bottom-up' initiative may thus be able to gain legal legitimacy by adopting the status of a 'non-profit' or charity dedicated to action for the public good, but this will not typically convey adequate standing and status in relation to planning and policy processes, and thus at least a measure of influence, if not decision-making, in such processes. Thus, higher-level government recognition may be required to convey legal legitimacy by statute: for example, via the creation of catchment or river basin authorities with responsibility to manage water resources as prescribed in the authorising statute (created by law and thus immediately acknowledged by the law). The HREP (Chapter 5) provides an example of a watershed organisation created by New York State statute that fosters good management through coordination, technical support and funding incentives. HREP is a unique technical and funding provider. Although it was established as a unit in the state's environmental regulatory agency, HREP serves only a planning and advisory function. It is a facilitator. It has no regulatory function and could not successfully meet its purposes if it were a regulator.

Legislatures are the ultimate policy-makers and their statutes are legally legitimate, binding and enforceable. They have the advantage of being fixed and accessible, and thus provide clarity on responsibilities and on actions permissible. Statutes have the force of an elected legislature behind them and, as the result of a democratic process, are generally well-accepted. However, statutes suffer the disadvantage that they frequently require a long time to enact, as the legislation process may be intentionally cumbersome and slow. The efficacy of an authority created by statute also depends on how it is monitored and enforced. If an organisation is not held accountable for fulfilling its statutory mandate, it may exist 'on the books' while becoming defunct in reality. The legal status conveyed by a statute can also depend on the prevailing hierarchy of laws which determines how much power an institution has. For example, in the United States federal laws are supreme, followed by state and then local laws. This legal hierarchy is important for small-scale catchment management organisations as their influence on decisions may be limited by this legal context despite their own legal status. For example, a local government authority such as DCAP (Chapter 4) may lack political recognition and higher-level influence within the region and state, despite its legal standing at the county level.

It is clear that there is no one model that fits all circumstances. Developing a legal framework for decentralised implementation of catchment management resembles a 'chicken-and-egg problem'. Traditionally the law is defined to codify existing social realities, its role being to discern legal relationships and obligations and organise them into systematic normative frameworks that facilitate communication, cooperation, action and enforcement. Challenges arise insofar as integrated, collaborative and adaptive catchment-based management, led and coordinated at the local level, has only relatively recently been recognised as a necessary approach for water resources protection and management and thus lacks the 'concrete reality' needed to inform legislation.

At the same time the complexity, lack of certainty and need for location-specific 'customisation' inherent to catchment management make it ill-advised for legislators to attempt to anticipate what legal relationships and obligations in such a reality should look like.

Answers to this dilemma may lie in case experience as illustrated by the innovative examples presented in this book. In particular, experience suggests that when available in a given jurisdiction a Memorandum of Agreement (MOA) or a Memorandum of Understanding (MOU) may be the best available legal mechanism to use to help the formative stages of a catchment management programme. A MOA/MOU is a legal instrument setting forth an arrangement to which the signatories agree. Its legal enforceability may be somewhat unproven and this may vary by jurisdiction, but it is a very flexible tool through which parties can agree to work in the manner which they desire and as appropriate to their situation. Unlike inter-municipal agreements, MOAs often have numerous signatories and can create new entities through which the signatories conduct their affairs. MOAs can thus be an effective means for parties to formalise and enhance the legitimacy of their agreement to work together to manage the land and water resources of a catchment, providing a crucial mechanism to enhance horizontal and vertical coordination, status and accountability, when existing multi-level governance arrangements lack provision for this.

The legal enforceability of MOA/MOUs remains uncertain because they do not have a lengthy history in the court system and will be treated differently in different jurisdictions. Typically, they can be expected to have less enforceability in court than contracts, which have an abundance of common law on which to base determinations of their efficacy. In addition, while MOA/MOUs provide a means to commence formalisation of collaborative catchment management activities, and can thus be a key means of protecting and managing water resources, they do not have any legislative capacity and thus cannot formulate policy or make decisions affecting the public. Their effectiveness is found in creating partnerships for catchment management, and in developing strategies that can be adopted by local and higher-level authorities such as municipal, county, and state or provincial governments. MOA/MOUs are therefore a potentially important tool for formation of collaborative catchment management, and for continuation of partnerships and networks. Specific regulatory responsibilities, for example permitting for water abstraction, remain with the agencies concerned. Thus the MOA or MOU employed does not become a means for formal governance of water resources.

## In conclusion

With reference to our case study examples, this chapter has considered why and how catchment management programmes start. It has revealed complexity and diversity in the drivers of action and in the processes of initiation and programme formation. The lessons that emerge reinforce our understanding that the characteristics of water resource management challenges, the management functions of, and the legal provisions for catchment management, all necessitate delegation to, and acceptance of responsibilities by, relevant local authorities and agencies. Catchment management also necessitates an integrated and holistic approach, and one that requires horizontally coordinated collaboration and partnership working between the organisations and stakeholder groups concerned at a catchment or sub-catchment scale.

These delegated governance arrangements must also depend on higher-level direction and authorities. Vertical coordination with national or regional policy objectives, legal and regulatory authorities, and funding, requires effective communication and interaction between higher and lower levels of government. The chapter identifies that this raises a number of dilemmas in terms of governance arrangements. To achieve its goals and sustain the participation and commitment of its stakeholders a catchment management programme and the body that coordinates it, require sufficient status, standing and legitimacy. Yet the codification of governance arrangements for this is very difficult to 'blueprint' for any given jurisdiction, and will often need to evolve with action and over time. There is also a need to balance the need for standing and legitimacy (and thus effective decision-making influence if not power) with the need to maintain independence, broad-based acceptance and trust as a facilitator and technical provider. Our leading example in this respect is that of the USC (Chapter 3), a partnership organisation that is a facilitator and technical provider but decidedly not a regulator. It is an organisation that has standing confirmed and legitimised by county governments, and influence with higher-level government that is achieved through demonstration of technical competence and performance as a cost-effective delivery mechanism. This does not, however, reduce or avoid the frustration and challenges that can arise when higher-level government imposes objectives and regulation poorly matched to local conditions (as exemplified by the example of the Chesapeake Bay Program watershed total maximum daily load, TMDL, targets set for New York State – see Chapter 3).

# 11 Getting informed

## Tools and approaches for assessment, planning and management

*Laurence Smith, Kevin Hiscock, Keith Porter, Tobias Krueger and David Benson*

## Introduction

Chapters 1 and 2 of this book argued that the technical and social complexities and uncertainties of catchment management, the potential for conflict between interest groups and need to resolve tradeoffs, and the need to establish an equitable and accepted sharing of costs and benefits all lead to the necessity that a catchment management programme has solid scientific credentials, such that its assessments and recommendations stand up to technical scrutiny, and have credibility with all stakeholders.

It was further proposed that the uncertainty, complexity and dynamics of catchment management challenges necessitate an adaptive management approach which is commonly understood as an iterative process of assessing system performance, implementing management actions, and monitoring and learning from outcomes. It is an approach that through processes of 'social learning' can incorporate co-creation of knowledge with stakeholders, change stakeholder perceptions, values and behaviours over time, and strengthen analytical capacity, social capital and trust between partners. For this process to occur it is necessary that technical information and scientific knowledge can be adequately accessible to, and shared by, all participants.

The relevant key question identified in Chapter 2 is as follows:

**3    What approaches and tools are needed for catchment management?**
(Which approaches and methods work? How can science be integrated into effective catchment management?)

In this chapter, we therefore reflect on how science and governance can be integrated into catchment management to achieve a technically sound, credible and accepted basis for planning and decision-making, and how an adaptive approach incorporating social learning can be effectively applied. The case studies in this book provide illustrations of these essential processes within catchment management.

## Which approaches and methods work in catchment management?

In organising this chapter, it is difficult to distinguish clearly between 'tools' and 'approaches' best employed in catchment management. This is because the two work in iterative combination, in ways that are themselves an integral part of the collaborative, adaptive and holistic approach of these programmes. However, in an attempt to reduce successful catchment management programmes to some of their key component parts, the next two sections consider the following. First, components labelled as 'tools' are decision-support tools, monitoring and communications. Second, components labelled as 'approaches' are analytic-deliberative processes and specific organisational and institutional provisions for the integration of science and governance.

The ways in which these tools and approaches iteratively combine in catchment management is demonstrated in microcosm by the development of the Loweswater Care Project (LCP; Chapter 9). From the earliest stages of formation of the LCP it was recognised that new forms of partnership working, re-framing of perceived problems and solutions, and co-production of knowledge were needed in response to frustration with past failure to solve the problem of algal blooms in the Loweswater lake. This situation engendered an inclusive and holistic approach, with the water quality issue becoming set within the 'problem' of integrated catchment management and encompassing all human activity, together with the catchment's ecosystem functioning. This need required communication between parties, including the effective transfer of technical information by scientists to 'lay stakeholders'. It also required the building of trust, particularly where relationships had been strained in the past. Participation by stakeholders from farmers to government agencies became inclusive and sustained. As confidence and knowledge of water quality problems grew, communication then extended to public outreach and awareness-raising. Knowledge-sharing and influence hence occurred both horizontally (with other areas and groups in the region) and vertically (with higher levels of government).

Lake monitoring data were downloaded by telemetry from sensors mounted on a lake buoy, and along with other monitoring data were made live and publicly available by upload to the project website. A mix of local stakeholders and professional researchers worked together to use such data to assess problems and identify solutions. For example, the formulation and use of linked modelling approaches as decision-support tools (Chapter 9) incorporated a wide range of expertise from local land managers to scientists. Local engagement with monitoring helped improve understanding of the causes of the water quality problem through provision of more accurate location-specific information, evaluation of management scenarios and the selection of solutions accepted and understood by the community. Linked models were used to understand land management impacts on water quality, providing a means to explain to non-experts how nutrient loads in the lake were linked to runoff, farm management practices, homes in the catchment and to the algal blooms. All of the scientific investigations conducted in the project were communicated and reviewed at LCP meetings. Community understanding and acceptance of the legitimacy of the work was essential for its use by the LCP in informing deliberation and planning.

## Catchment management tools

Decision-support tools developed through computer-based modelling, and combined with monitoring and communication, can serve a number of purposes related to the protection and management of water resources. Used together, monitoring and modelling can inform planning and decision-making by improving understanding of how environmental systems work, and how they may react to changes made to land and water management. Models combine and reconcile data with process knowledge and theory in a way that allows 'if–then' scenarios to be developed and tested. Models can also provide a common learning platform through use of which stakeholders can share and integrate their diverse knowledge and perspectives. Use and location-specific application of models through co-development by scientists and other stakeholders can provide a formalised means to capture and preserve such emerging knowledge. Models can range from those that are simplistic and indicative, relying only on modest data inputs, to process-based models that are complex with correspondingly demanding data requirements.

Monitoring requires considerable time and resources and thus in most catchments can only be conducted on a selective and limited basis. Use of modelling to simulate natural systems and provide predictions of outcomes at a given site under specified conditions can thus be used as an alternative to comprehensive monitoring. However, models require data, and thus dependent on conditions and objectives a minimum monitoring requirement will remain. For example, location-specific monitoring at a small sub-catchment scale can be used to inform the construction and validation of models, the use and interpretation of which must remain cognisant of the inherent data limitations and uncertainties.

### Decision-support tools

Development of a catchment management programme requires effective means to understand or characterise the natural environment of the catchment, and to assess the pressures and threats to water resources that exist. For example, the 'streams action agenda' of the HREP (Chapter 5) promotes a locally led approach to catchment planning and conservation that includes: preparation of an inventory of catchment natural resources; identification of the source and degree of threats to these resources; and priority setting for protection and restoration.

Non-point sources of pollution are a major threat to water quality in most catchments, but quantifying these, and the benefits of management measures, can be difficult and demanding of data. Under the DCAP (Chapter 4) data from water quality monitoring stations and other field studies in the Cannonsville Reservoir basin were used to support comprehensive mathematical and statistical analyses and mathematical modelling by scientists at Cornell University that subsequently informed management decision-making and helped evaluate actions implemented in the catchment. The USC (Chapter 3) similarly makes use of modelling to understand nutrient and sediment fluxes throughout the Upper Susquehanna Basin, and to plan implementation of control measures based on appraisal of their potential effectiveness.

In South East Queensland, the Healthy Waterways attempts to reflexively respond to changes in environmental and socio-economic contexts within its catchments. To enable this approach to work effectively, a number of decision-support tools have been developed including the 'Source Catchments' model (Chapter 6). This is an integrated water quality and quantity model designed to help programme managers develop targets, prioritise improvements and measure the effectiveness of a broad range of catchment management activities. In this respect, it models the amounts of water and contaminants flowing through a catchment and into receiving water bodies. The model can also be used to predict the flow and pollutant loads at any location in a catchment over time. Users can test different scenarios including changes in land use, land management and climate, helping to answer management questions such as where to place measures to optimise water quality (eWater, 2014).

A key premise of a collaborative approach to catchment management that we derive from the case studies in this book is that stakeholders with differing understandings and values can deliberate on management options and develop shared plans. This approach requires a shared knowledge base and development of common understandings of catchment bio-physical processes. In catchment science, as in engineering, the application of computer simulations is attractive because they may enable prediction and evaluation for problems that are often difficult, costly or impossible to observe directly. Such catchment science problems can include, for example, the morphological and ecological impacts of engineered structures, flood risk, soil and groundwater processes that condition pollutant behaviour, climatic change, and crop yield response to conditions such as drought and reduced fertiliser application.

Modelling enables 'what if' scenarios to be evaluated, at least within bounds set by the areas and subject of the investigation, thereby reconciling scientific theory with field empirical evidence. Models may thus inform where and what new field investigations should be undertaken to reduce predictive uncertainty (Krueger et al., 2007). In many situations modelling may interpolate between spot field investigations and measurements that may be too difficult and costly to repeat everywhere in a catchment. It is important to note, however, that all models need data to parameterise, drive and test them, so investment in modelling necessitates a commitment to maintaining at least minimum essential monitoring networks.

As illustrated by the examples from this book and summarised above, the use of decision-support tools in the form of computerised models provides a means to communicate understanding, and enable a framing of the scale and severity of water management problems with stakeholders. Models can be used to explore the potential effectiveness and tradeoffs of alternative pollution control measures and management scenarios, a process that will usually stimulate highly dynamic and engaged debate. This process of analysis and debate can then progress to the costing of management options and a collective assessment of the capacity of existing organisational and institutional arrangements to deliver chosen solutions.

Point and non-point sources of nitrate and phosphorus pollution are leading threats to water quality in most catchments (Chapter 1). Consequently the identification and quantification of sources, and in particular delineation of critical source areas that contribute disproportionate amounts of sediment and nutrients to receiving waters (e.g. Heathwaite *et al.*, 2005; White *et al.*, 2009) has been a prime focus for development of models[1] (Bouraoui and Grizzetti, 2014). Prioritising and targeting diffuse pollution control measures according to expected impacts and associated costs, through use of models to identify and rank critical source areas, has the potential to optimise investment in pollutant reduction, including minimisation of the extent of land affected by economically restrictive land use practices (Bouraoui and Grizzetti, 2014). Such optimisation is important given often limited financial resources, and the need to limit the economic and social costs of meeting standards such as those set by the USA Clean Water Act or EU WFD which might otherwise discourage full adoption and implementation of the legislation (Bouraoui and Grizzetti, 2014).

The use of models for this purpose does, however, remain a demanding challenge and an imprecise science. Assessment of the cost-effectiveness of measures to combat diffuse pollution is difficult because of the spatial and temporal lags that usually occur between taking a management action at a local scale and the outcome for water quality. Numerical modelling does not pretend to be absolutely precise in its predictions, at least in the environmental sciences, on account of the complexity of the real world and (often) paucity of input data. To be credible, vital aspects of model application are 'calibration' (the setting of parameters at data input, usually such that model outputs compare with their measured equivalents) and 'validation' (the comparison of calibrated model outputs with independent data). Calibration and validation will never be perfect, hence it has become 'state-of-the-art' to attach some measure of uncertainty to model parameters and subsequently predictions (Krueger *et al.*, 2007).

Model selection or construction should be based upon considerations agreed by stakeholders participating in the model production process. Essentially this is the establishment of a 'conceptual model' that qualitatively describes real-world processes and is based in good science and observation. The model operation, at least in a qualitative sense, should also be understandable for the team of experts and stakeholder partners working in the field. There should be sufficient data (in terms of range, accuracy and volume) available as input, and there should be sufficient historic, or independently gathered, data for model validation (i.e. the same data should not be used in the modelling as is employed in assessing the probable accuracy of outputs).

Meeting these conditions in model development and application is important in gaining acceptance of model results by diverse stakeholders, and thus important, for example, in gaining acceptance and adoption of proposed diffuse pollution control measures. As illustrated by several of the case studies in this book, models have proven to be effective decision-support tools when used to communicate the predicted impacts of management measures and involve stakeholders in the discussion of alternatives. A key factor is the involvement of concerned stakeholders from the earliest stages of model use as this inclusivity can improve the formulation, adoption and implementation of pollution mitigation plans.

Yet the use of models should not be seen only as a means for explaining and illustrating how catchment processes work, and for involving stakeholders in planning and decision-making. An increasing number of models for catchment assessment and management depend for their development and application on the incorporation of 'expert opinion' sourced from the diverse experience and knowledge of stakeholders (Krueger *et al.*, 2012). The development and use of such models goes 'hand in hand' with adoption of participatory and adaptive approaches to the complexities of catchment management. The aim is the co-production of models by a diverse community of experts and stakeholders, working in partnership to achieve shared goals.

The benefits of participation have been categorised as substantive, instrumental and normative (Fiorino, 1990), and this is well-illustrated by stakeholder co-production of catchment management models. The substantive argument holds that participation can capture more effectively all the available knowledge and perspectives necessary to manage complex social-ecological systems and thus lead to better outcomes. For example, substantive benefits from catchment modelling arise because conclusive measured data may never be attainable at an affordable cost and in due time for some parameters, given the complexity and scale of environmental systems. This often leaves local expert opinion as the only source of evidence (Vrana *et al.*, 2012). This situation is likely to be most evident when timescales are short and policy demands urgent so that expert opinion is needed to bridge data gaps and operationalise models. The integration of scientific with location-based (local) knowledge can also simply lead to better models and better decisions.

The instrumental argument for including such evidence implies more broad-based acceptance of outcomes through participation and thus easier conflict resolution and implementation of policies. Co-production and use of models can be a form of social learning by a stakeholder community in partnership with scientists that leads to greater trust in the analysis and decisions, reduced conflict, and stronger commitment to implementation through a sense of ownership. Again the Loweswater case provides a leading example as described in Chapter 9, and in more detail for its modelling by Norton *et al.* (2012).

Lastly, these pragmatic substantive and instrumental gains from stakeholder participation in model development and use are complemented by the normative arguments that individuals have a right to participate in science and policy processes that result in outcomes that affect them (Lafferty and Meadowcroft, 1996; Tsouvalis and Waterton, 2012), a rationale that draws on a wider literature on deliberative democracy (e.g. Freire, 1970; Habermas, 1984).

In practice, most catchment management programmes will tend to select from existing models for application because model creation and development is both specialist and time-consuming. Where expertise is available geographic information system (GIS) applications will often be used as a means of capturing, manipulating and displaying spatial data, such as vulnerability mapping and the outputs of pollution source identification and apportionment models. GIS tools can be particularly useful in reviewing and presenting both the current understanding of the pressures and problems in a catchment and the current and proposed interventions of organisations responsible for improvement measures.

No single 'off-the-shelf' decision-support tool or model is likely to provide all the solutions needed for a given catchment. Indeed, as all models rely on assumptions and input parameter estimates as well as measured values subject to error and uncertainty, any single model's output must be regarded as an estimate and subject to an error margin. Some programmes may thus choose to apply alternative models to test their focus and application, or indeed rely on aggregating a range of results from multiple models.

Different models have different objectives and thus there may be benefits from using models in combination in a stepped or tiered approach. This process can provide a means to exploit the complementarity of different tools in assessing a range of problems and solutions as iterative planning for management develops, taking account of interplay between the scale of application, purpose, data needs and accuracy of different tools. For example, for a given catchment with a water quality problem the modelling may be approached as follows.

1   **Source apportionment at whole catchment scale.** It will usually be important and useful to establish at an early stage of assessment whether a catchment is dominated by point source or non-point sources of pollution, or by a mixture of both. As noted in Chapter 1, mitigation options for each of these are very different in scope, potential costs and difficulty. Point/non-point source separation and apportionment may be achieved by statistical analysis of available water quality monitoring data (e.g. Bowes *et al.*, 2008).

2   **Appraisal of scale and severity of problems at sub-catchment scale.** A second step, particularly to focus on non-point sources of pollution and possible mitigation measures, will need to be a more nuanced simulation of the mix of pollution pressures and effects of catchment

interventions. Working at the sub-catchment scale, and corresponding level of detail, will help stakeholders relate to the issues and facilitate a meaningful discussion of possible solutions with the whole catchment in view. This level of analysis may be provided by an export coefficient type model (e.g. Zobrist and Reichert, 2006).

3  **Field scale targeting of mitigation efforts.** Where true diffuse or non-point sources of pollution are the focus of concern, a third necessary step is to identify vulnerable and critical source areas of land and match these with mitigation options. As the focus here is on mapping, the demand that the model simulates water quality outcomes can be relaxed in return for a finer spatial detail that would not support a process-based model. Water quality outcomes can still be simulated at the sub-catchment scale (step 2). Suitable mapping tools are spatial risk-based models that are only constrained in resolution by the available digital topographic and other thematic maps (e.g. Reaney *et al.*, 2011).

4  **Farm-scale cost–benefit appraisal of mitigation measures.** As soon as potential agricultural and other land management pollution mitigation measures are identified for specific locations, the fourth step should be a thorough appraisal of feasibility, acceptance and cost–benefit outcomes for the individual farms involved (e.g. Zhang *et al.*, 2012).

*Monitoring and communications*

Monitoring provides the means to collect information about how systems naturally operate. Data collected by monitoring can be used to quantify and understand processes (and as noted above it can be used to conceptualise, construct, calibrate and validate models), or to identify the source of water quality problems, both leading to identification of potential management options. Monitoring can also be used to establish baselines against which future trends can be identified and measured. Thus, monitoring provides the means to evaluate any changes in water quality resulting from change in management practices. Lessons learnt from such evaluation form the basis for an adaptive management approach so that iterative improvements can be made over time. Similarly, the cost-effectiveness of actions taken can also potentially be 'fine-tuned' through successive implementation of action plans.

In thinking about the uses and benefits of decision-support tools, the communications value of models, when used as part of effective monitoring and reporting, should not be underestimated. Although not a computerised model, a leading example of a decision-support and communications tool cited in this book is the annual Report Card (backed by a full technical report) used by the Healthy Waterways in Queensland (Chapter 6). This has been published each year since 1999 to summarise and communicate information on the ecosystem health of the region's waterways. It consists of a data table and a map-based graphic and serves to:

- provide an understandable 'snapshot' of the health of the region's fresh-water, estuarine and marine environments in relation to environmental targets and standards;
- raise awareness of change in the condition of waterways over time and the success of local action programmes, thereby building understanding of the effectiveness of improvements in land and water management;
- act as a form of model to help focus management efforts and resource allocation by partner organisations to protect and improve vulnerable and degraded areas, and to demonstrate and communicate the predicted impacts of possible future scenarios and management plans.

Most importantly, the Healthy Waterways Report Card should also be recognised as a political tool; and a prime example of how science integrates with governance. The publication of accessible information about outcomes supports the accountability and legitimacy of those responsible for taking action, and can influence the resource allocations made by local and higher-level government partners. It similarly encourages participation by industry and other civil society groups.

Most decision-support tools, and particularly those with attractive and accessible user interfaces, can utilise monitoring data in a reporting and communi-cations role, as well as be used in their conventional planning and appraisal role. Effective communication of scientific information, and particularly of outcomes, is essential to the continued and sustained dynamic of a successful adaptive management approach. These tools can serve to reward and reinforce participation by stakeholders by informing them of the effectiveness of their contributions, and influence decision-makers to continue commitment of political support and funding for a catchment management programme.

For example, Chapter 5 noted how in the HREP it is recognised that the scientific information gathered by the many studies conducted by the programme's partners needs to be interpreted and explained in a format accessible to the public. This enables stakeholders, including members of the general public, to monitor progress for themselves, evaluate the effectiveness of programme activities and participate in ongoing decision-making or advocacy and lobbying.

Chapter 9 further illustrated how such processes can function and be an important and effective mechanism for change at a very local level. For example, LCP meetings were advertised through the local parish newsletter, invitation cards for meetings were widely distributed in the locality, and accounts of meetings, activities and achievements were published via the parish newsletter and 'community noticeboard' of the project website.

Common communication problems that can arise between scientists, resource managers and other stakeholders include use of concepts and terminology, and variation in priorities and understanding of which scientific findings are most relevant to the issue at hand. One tool that has been recognised to improve information exchange is a central, shared and accessible database comprised of reliable information that combines scientific and local expert knowledge. For such a resource, attention has to be paid to issues surrounding the comparability of data, data records (metadata) and any restrictions on data use and access, although internet-based tools do increasingly provide the means to overcome these constraints (UNU-INWEH, 2012).

### Catchment management approaches

In this section, we consider why sound scientific assessment and stakeholder deliberation are a necessary duality at the core of integrated, collaborative and adaptive catchment management programmes, and we identify from the case studies in this book some specific organisation and institutional provisions that aim to achieve this synergy.

#### Adaptive and analytic-deliberative processes

Chapter 1 introduced the challenges of controlling diffuse water pollution, and in particular its inherent scientific and societal uncertainties. The first of these difficulties relates to technical uncertainties regarding pollution sources, pathways and impacts, and to the efficacy of control measures and governance arrangements; the second to variation in people's values, preferences and willingness to bear the costs of water resources protection and improvement. Such uncertainties and complexity were presented as key features of catchment management characterised as a complex or even a 'wicked problem'.

This characterisation of catchment management as a difficult environmental problem characterised by goal conflict and uncertainty suggests that the development and selection of measures to protect and conserve water resources needs to be achieved through a process of constant improvement of scientific understanding and technical management of the catchment, conditioned by parallel deliberations concerning priorities and resource allocation by stakeholders and mandated decision-makers. Thus, a key premise is that a twin-track approach of scientific research and stakeholder deliberation is needed. The potential substantive, instrumental and normative benefits of participatory approaches were outlined above. Local expert opinion can provide a form of applied peer review for conventional science: potentially exposing biases, inconsistencies and oversights; redefining boundaries in socially relevant ways; questioning scientific assumptions and inference rules; and adapting standard solutions to local conditions.

Understanding of the duality of scientific and societal uncertainty leads to recognition of the need for a twin-track or analytic-deliberative approach of applied scientific research and stakeholder participation and deliberation. Thus research to improve understanding of bio-physical and socio-economic processes in catchments should be integrated with a deliberative process situated at the appropriate scale and level of governance. It is important that the deliberative process should drive the scientific agenda, recognising that actions should be the outcome of decisions determined by the values and choices of the people affected and informed by scientific data and technical understanding. In such a process, stakeholder deliberation can formulate and legitimise action plans at a local scale, and it can improve the quality of professional enquiry through the integration of independent scientific and local contextual knowledge (as discussed above with regard to the development and application of decision-support tools). Although the case for stakeholder participation in environmental management is widely accepted (e.g. Creighton, 2005; Fischer, 2000), this duality of research and deliberation is rarely made sufficiently explicit. As a consequence, inadequate attention has generally been paid to the processes, institutions and resources that help it function effectively (Smith and Porter, 2010).[2]

All of the case studies in this book illustrate the centrality of this adaptive and analytic-deliberative approach in catchment management. A leading example is provided by Healthy Waterways in Queensland. As the capacity of the partnership and its integrated approach to water management evolved over time, as described in Chapter 6, priority shifted from an initial focus on technical studies used to characterise problems, generate data and develop decision-support tools, to the development of collaboration and partnership working based on the agreement of a common vision and the setting of goals. The stated and strategic aims of the partnership became underpinned by two principles:

- a commitment to working in a coordinated partnership structure in which all partners can be heard, contribute to decision-making and implement agreed actions within their own spheres of responsibility;
- formulation of management strategies on the basis of sound science, rigorous monitoring of the waterways environment, and adaptive learning.

These principles exemplify a twin-track or analytic-deliberative approach, as made manifest by the investments made in decision-support and communication tools, and in processes of knowledge co-generation with stakeholders. Healthy Waterways has invested heavily in both the natural and social science of water management (understanding, modelling, monitoring and evaluation) and the collaborative generation of knowledge with stakeholders as a basis for learning and adaptive management.

The three examples of watershed management in New York State similarly provide examples of an analytic-deliberative approach. The stated mission of the USC (Chapter 3) 'is to protect and improve water quality and natural resources in the Upper Susquehanna River Basin with the involvement of citizens and agencies through education, partnerships, planning, implementation and advocating for our water resources' (USC, 2014). By addressing issues (e.g. streambank erosion) at the source (e.g. headwaters), across stream corridors and the landscape, and by addressing them programmatically (e.g. through education and local involvement), the multiple barrier approach that the USC adopts for watershed planning and management captures stakeholder interests by incorporating their priorities and local knowledge, and by demonstrating progress through implementation.

The DCAP in the New York City Watershed (Chapter 4) provides another leading example of a comprehensive management programme for watershed protection based on scientific credentials but delivered under local auspices. Its establishment represents a novel attempt to protect water resources, based on land management, through scientifically and economically sound governance based on local democracy (Porter *et al.*, 2006).

As was explained in more detail in Chapter 5 the HREP is guided by a forward-looking 'Action Agenda' developed through significant and sustained community participation along the length of the river. Building on scientific understanding and principles of ecosystem-based management, the programme is steered by an advisory committee, the broad-based membership of which helps engage many representatives of partner organisations and the public in working together toward common goals (NYSDEC, 2014). Grant funding of projects has allowed counties, towns and villages in the Hudson River Valley to take ownership of their resources and define the future of their communities and environment, with scientific and technical support.

Chapter 7 demonstrates that planning and action in each of the three cases of groundwater protection was also underpinned by scientific assessments that provided a conceptual and practical understanding of the local and regional groundwater system in the context of its soils, hydrogeology and water body characteristics. To minimise costs in terms of lost output from land use and to optimise water protection, measures need to be specific and well-adapted for local soil types and geology. This requires a comprehensive and accurate technical understanding of a groundwater catchment area and its underlying aquifer. Thus essential capabilities for management of a groundwater protection programme are monitoring and evaluation of water quality and quantity, and identification and analysis of pollution sources and threats. Capability is also required for the preparation and implementation of spatial plans at a local or regional level. In the three cases in Chapter 7 this technical knowledge underpinned identification and implementation of solutions from field-scale farm management practices to integrated landscape-scale spatial plans. Critically, local acceptance and successful implementation was only achieved through the information being made widely available, shared, debated and trusted. Changing land use management or the land use pattern in an area can be a sensitive issue subject to challenge, and may require sufficient time for an extended process of negotiation. The experience reported in Chapter 7 suggests that communication and the coordination of change must be based on trust and continuity of information-sharing and action. It was also found to be important that independent people lead the central liaison role and not government or a regulatory body.

For the WWF RIPPLE project, Chapter 8 describes how the project team embarked on community-led participatory planning through a carefully designed stakeholder engagement process, involving multiple stages and multiple actors. It should be noted that it took 24 months from project inception to arrival at an agreed catchment management plan. This lengthy timeline reflects the intensive level of interaction which took place with the community and the iterative nature of approaches that again demonstrate the necessary duality of technical assessment and local deliberation.

*Specific organisational and institutional provisions for the integration of science and governance*

The societal uncertainty involved in water resource management has been described above. Catchment management inevitably involves issues which are contested and over which there may be conflict of interests. It is thus essential that a catchment management programme should have solid scientific credentials, and that its assessments and recommendations stand up to technical scrutiny and have credibility with all stakeholders. This feature underpins the analytic-deliberative approach illustrated by the case study examples in this book, but is not something that necessarily happens automatically. What governance provisions can therefore be made for this central and key mechanism to work effectively?

Perhaps the most explicit answer to this question is again provided by Healthy Waterways (Chapter 6). The need to establish the relevant science and characterise the environment was recognised early in the partnership's development, not least because a lack of scientific understanding of the impacts of pollution was thought to provide an excuse for political inaction. Experience then showed that robust scientific research, modelling, monitoring and communication all facilitated the engagement of partners and other stakeholders, integration of their needs and priorities into action plans, and the moderation of erroneous and unhelpful beliefs and attitudes. The earliest and most vital organisational and institutional provision that was made was the formation of the Scientific Expert Panel (SEP), with membership drawn from local and national research organisations and key partners. The SEP has demonstrated sustained importance, value and influence through its role in ensuring that the Healthy Waterways' scientific programmes and decision-support tools have local applicability and credibility.

The essential requirement for actions to be based on sound scientific credentials was also an explicit, core and foundational principle for the DCAP in the New York City Watershed (Chapter 4). Provision for this purpose was made for DCAP through the creation of its inter-agency Scientific Support Group (SSG). This body has the goals of:

- provision of a scientific basis for assessing management needs and options in the watershed;
- verification of scientific validity of management measures;
- evaluation of management options implemented (Porter *et al.*, 2006).

Chapter 3 similarly explains how both the USC's action programmes, and its contributions to strategic planning for the Chesapeake Bay Program, are underpinned by strong 'in-house' scientific capability supported by research partnerships with the leading universities of the region. Such partnerships do not function effectively overnight, but again require specific intent and provision to develop organisational lines of communication, linkages and institutional arrangements that facilitate funding and implementation. Likewise the scientific foundation of the HREP (Chapter 5) has been strengthened through collaboration with regional academic and research institutions that aim to make the HREP a model and an exemplar for scientific management of a major watershed implemented through diverse and varied partnerships.

In contrast, the WWF RIPPLE project (Chapter 8) is a community-led project rather than a science-led initiative, but one that nonetheless still utilises scientific information and expertise when needed to help its champions and delivery groups in planning and implementation. A particular feature of the project's provision has been allowance for a leading role to be taken by community volunteers in obtaining scientific data directly themselves, or with the assistance of the project coordinator. First initiated as a 'speed-dating' exercise between project champions and the project advisory group, this form of data generation became a continuous activity.

The LCP was cited at length as an exemplar at the beginning of this chapter. One further innovative aspect of the research project associated with this catchment programme that is worth highlighting, was the modest financial provision for the community organisation (LCP) to commission its own research studies. This mechanism was considered to act as a very effective 'leveller' between community and scientists, and a facilitator of community engagement. It also generated robust technical information of very direct relevance to the management needs of the catchment.

## Conclusions: How best can science be integrated with governance for effective catchment management?

The case studies in this book suggest that it is valuable to have high-quality informational tools and decision-support systems to communicate technical information to informed lay stakeholders: tools that can effectively integrate measured empirical data with location-specific details that local people and partner organisations are able to provide but also consequently support. Where trust and a shared vision has been established, such tools can be employed in participatory and deliberative processes that underpin an analytical-deliberative and adaptive management cycle.

The experiences cited in this book encourage us to advocate the use of models as effective tools for identification and apportionment of pollution loads, appraisal of the cost-effectiveness of mitigation measures, targeting of such measures in terms of location and timing, and as a basis for involving stakeholders in co-production of the tools and analyses necessary for appraisal of options and development of management plans. Used singly or in combination, models are now available that allow assessment of alternative management options and can guide stakeholders in determining sustainable pollution loadings and achievable water quality targets for the catchment. When supported by the best available peer-reviewed science, which demands quantification and communication of scientific uncertainty, this planning process provides a credible and sound basis for implementation.

Monitoring of actual systems behaviour and outcomes provides the crucial means to close the adaptive management cycle, so that lesson learning improves future planning and implementation. However, resources for monitoring are inevitably limited and their allocation can be best guided by the same analytic-deliberative process underpinned by sound science, and well-established institutional mechanisms to provide credible scientific review and verification.

Yet, all of these processes depend on effective communication of technical information. There is a need to communicate information about catchment problems and solutions in ways that are relevant to stakeholders, understandable and actionable. Communication should be understood as a means to promote inclusiveness, a means to raise awareness, educate and influence behaviour, and as a political means for advocacy and the securing of resources.

## Notes

1   For example, Bouraoui and Grizzetti (2014) review a diverse range of models capable of simulating the impact of nitrogen mitigation measures, 'ranging from empirical methodologies (EXPORT Coefficient) and conceptual models (GREEN, MONERIS, MITERRA) to physically-based models (HBV-N/HYPE, STICS-MODCOU, EPICgrid, SWAT, INCA)'.

2   A leading exception is provided by Innes and Booher (2010) who explore transdisciplinarity understood as integration across scientific disciplines and other (citizen) strands of knowledge. They emphasise processes of permanent collaboration between public and private decision-makers with scientists acting as informants and facilitators throughout a whole planning and decision-making process. Rational communication among stakeholders informed by science becomes the key tool to adaptive mitigation of wicked problems that are subject to contingent and ever-changing characteristics. Parallels for this process can also be seen in the 'analytic-deliberative' model for characterisation and communication of risk recommended by, for example, Stern and Fineberg (1996), Stern (2005) and Burgess *et al.* (2007). Further parallels can be seen in the notion of 'post-normal science', an approach considered appropriate for situations where 'facts are uncertain, values in dispute, stakes high and decisions urgent' (Funtowicz and Ravetz, 1991, p. 138). The 'analytic-deliberative duality' is evident in the expected practice of post-normal science in which stakeholders affected by an issue enter into dialogue about it, and bring to this dialogue local knowledge and experience.

## References

Bouraoui, F. and Grizzetti, B. (2014) 'Modelling mitigation options to reduce diffuse nitrogen water pollution from agriculture', *Science of the Total Environment*, vol. 468–469, pp. 1267–1277.

Bowes, M. J., Smith, J. T., Jarvie, H. P. and Neal, C. (2008) 'Modelling of phosphorus inputs to rivers from diffuse and point sources', *Science of the Total Environment*, vol. 395, no. 2–3, pp. 125–138.

Burgess, J., Stirling, A., Clark, J., Davies, G., Eames, M., Staley, K. and Williamson, S. (2007) 'Deliberative mapping: A novel analytic-deliberative methodology to support contested science policy decisions', *Public Understanding of Science*, vol. 16, pp. 299–322.

Creighton, J. (2005) 'What water managers need to know about public participation: One US practitioner's perspective', *Water Policy*, vol. 7, pp. 269–278.

eWater (2014) *Source Catchments: Realising Whole-of-catchment Water Management*, http://www.ewater.com.au/h2othinking/?q=2010/08/source-catchments-realising-whole-catchment-water-management, accessed 25 April 2014.

Fiorino, D. (1990) 'Citizen participation and environmental risk: A survey of institutional mechanisms', *Science, Technology, and Human Values*, vol. 15, no. 2, pp. 226–243.

Fischer, F. (2000) *Citizens, Experts, and the Environment*, Duke University Press, Durham.

Freire, P. (1970) *Pedagogy of the Oppressed*, Penguin, Harmondsworth.

Funtowicz, S. O. and Ravetz, J. R. (1991) 'A new scientific methodology for global environmental issues', in Costanza, R. (ed.) *Ecological Economics: The Science and Management of Sustainability*, Columbia University Press, New York, pp. 137–152.

Habermas, J. (1984) *The Theory of Communicative Action: Reason and the Rationalization of Society*, vol. 1. Beacon Press, Boston.

Heathwaite, A., Quinn, P. and Hewett, C. (2005) 'Modelling and managing critical source areas of diffuse pollution from agricultural land using flow connectivity simulation', *Journal of Hydrology*, vol. 304, pp. 446–461.

Innes, J. and Booher, D. (2010) *Planning with Complexity: An Introduction to Collaborative Rationality for Public Policy*, Routledge, London and New York.

Krueger, T., Freer, J., Quinton, J. and Macleod, C. (2007) 'Processes affecting transfer of sediment and colloids, with associated phosphorus, from intensively farmed grasslands: A critical note on modelling of phosphorus transfers', *Hydrological Processes*, vol. 21, no. 4, pp. 557–562.

Krueger, T., Page, T., Hubacek, K., Smith, L. and Hiscock, K. (2012) 'The role of expert opinion in environmental modelling', *Environmental Modelling and Software*, vol. 36, pp. 4–18.

Lafferty, W. and Meadowcroft, J. (1996) 'Democracy and the environment: Prospects for greater congruence', in Lafferty, W. and Meadowcroft, J. (eds) *Democracy and the Environment: Problems and Prospects*, Edward Elgar, Cheltenham, pp. 256–272.

Norton, L., Elliott, J., Maberly, S. and May, L. (2012) 'Using models to bridge the gap between land use and algal blooms: An example from the Loweswater catchment, UK', *Environmental Modelling and Software*, vol. 36, pp. 64–75.

NYSDEC (2014) *Hudson River Estuary Program*, New York State Department of Environmental Conservation, www.dec.ny.gov/lands/4920.html, accessed 16 May 2014.

Porter, M. J., Porter, K. and Frazier, D. (2006) *Delaware County Action Plan: Linking Economic Vitality, Water Quality, and Sound Science*, Progress Report, Department of Watershed Affairs, Delhi, NY.

Reaney, S. M., Lane, S. N., Heathwaite, A. L. and Dugdale, L. J. (2011) 'Risk-based modelling of diffuse land use impacts from rural landscapes upon salmonid fry abundance', *Ecological Modelling*, vol. 222, no. 4, pp. 1016–1029.

Smith, L. E. D. and Porter, K. S. (2010) 'Management of catchments for the protection of water resources: Drawing on the New York City Watershed experience', *Regional Environmental Change*, vol. 10, no. 4, pp. 311–326.

Stern, P. (2005) 'Deliberative methods for understanding environmental systems', *Bioscience* vol. 55, no. 11, pp. 976–982.

Stern, P. and Fineberg, H. (1996) *Understanding Risk: Informing Decisions in a Democratic Society*, National Research Council, National Academy Press, Washington DC.

Tsouvalis, J. and Waterton, C. (2012) 'Building "participation" upon critique: The Loweswater Care Project, Cumbria, UK', *Environmental Modelling and Software*, vol. 36, pp. 111–121.

UNU-INWEH (2012) *Analysis Report of the Land-based Pollution Sources Working Group*, United Nations University Institute for Water, Environment and Health, The United Nations University, Hamilton, Ontario.

USC (2014) 'Welcome to the USC Homepage', www.u-s-c.org/html/index.htm, accessed 16 May 2014.

Vrana, I., Vaníček, J., Kovár, P., Brozek, J. and Aly, S. (2012) 'A fuzzy group agreement based approach for multiexpert decision making in environmental issues', *Environmental Modelling and Software*, vol. 36, pp. 49–63.

White, M., Storm, D., Busteed, P., Stoodley, S. and Phillips, S. (2009) 'Evaluating nonpoint source critical source area contributions at the watershed scale', *Journal of Environmental Quality*, vol. 38, no. 4, pp. 1654–1663.

Zhang, Y., Collins, A. L. and Gooday, R. D. (2012) 'Application of the FARMSCOPER tool for assessing agricultural diffuse pollution mitigation methods across the Hampshire Avon Demonstration Test Catchment, UK', *Environmental Science and Policy*, vol. 24, pp. 120–131.

Zobrist, J. and Reichert, P. (2006) 'Bayesian estimation of export coefficients from diffuse and point sources in Swiss watersheds', *Journal of Hydrology*, vol. 329, no. 1–2, pp. 207–223.

# 12 Getting things done and getting results

*Laurence Smith, Keith Porter and David Benson*

## Introduction

Chapter 2 sets out the complex and difficult challenges of water resource management that necessitate a catchment-based, collaborative and adaptive approach. Chapter 10 explores how such programmes start, and how they need to integrate 'bottom-up' aspirations and initiatives with the aims and regulations of higher-level governance. It was explained why effective catchment management programmes must be led and coordinated through acceptance of relevant responsibilities at local level, but also why inter-locality cooperation and coordination may be necessary at a larger spatial scale. Chapter 11 reviews the key methods and decision-support tools necessary to meet the information requirements of these approaches. It also explores the rationale for, and strengths of, inclusive stakeholder participation and an analytic-deliberative approach.

Adding to Chapters 10 and 11, this chapter reflects on how catchment management programmes mature and sustain themselves. The vision for catchment management that emerges is one of coordinated cross-sectoral partnership-working at a sub-catchment or whole-catchment scale, involving government, the private sector and civil society. Within such a partnership, delegations of responsibility for decision-making and action are well assimilated in multi-level governance arrangements, and are thus supported by, and remain accountable to, the relevant higher levels of authority. Processes of catchment assessment, planning and implementation by the partnership are enhanced and legitimised by knowledge sharing with, and participation by, other interested stakeholders. Delegated mandates within the partnership should not be unfunded. A combination of dedicated core provision, and/or resources from the existing budget lines of partners, can support capacity and action at local level, but successful programmes also demonstrate the ability to supplement this with funding gained from a diversity of sources (including grants and voluntary contributions in kind).

Such arrangements are variously described in the environmental management literature as collaborative, cooperative, polycentric, networked and multi-level. To offer and sustain a unified and integrated catchment management programme requires fostering of vertical and horizontal coordination, and an assimilation of duties and responsibilities in a partnership between higher and local government, government agencies, businesses, NGOs and other community organisations. This chapter seeks to illustrate how these arrangements, and the water resources management and other environmental and socio-economic outcomes that they deliver, can be sustained. As in earlier chapters, we turn to our case studies to provide empirically grounded examples of what can work well in practice.

The relevant key questions identified in Chapter 2 are as follows:

## 4   How do things get done?

(How can successful catchment management be kept going? What governance arrangements sustain effective catchment management?)

## 5   What are the outcomes of catchment management?

(How do we know that it works? What are appropriate measures of effectiveness and success? What capabilities must be demonstrated and other criteria met for a catchment management programme to be deemed successful and sustainable?)

## How to get things done: principles and the role of law

Chapter 10 identifies challenges relating to the formation of a catchment management programme and the establishment of sufficient legal status for the programme to be able to achieve its objectives through its standing with both local stakeholders and higher-level authorities. In terms of 'getting things done', key challenges for an ongoing catchment management programme are how to maintain that standing, and how to deliver the environmental and public health improvements that policy makers intend to bring about through their legislation and policy directives under the prevailing legal and governance framework.

The case studies in this book provide leading examples of attempts to implement national (or federal) policy goals as expressed in the law. For example, the federal Clean Water and Safe Drinking Water Acts in the USA, the Water Framework Directive (WFD) in the European Union (EU), and the National Water Quality Management and Natural Heritage Trust Strategies in Australia. Cases in this book also include examples of locally led catchment initiatives that are reciprocal to such national policies and programmes, for example, the Upper Susquehanna Coalition described in Chapter 3, the cases of groundwater protection in Chapter 7, the WWF RIPPLE project in Chapter 8 and the Loweswater Care Project in Chapter 9.

The organisational and institutional arrangements of a catchment management programme need to be legally legitimate, politically accountable, transparent, participatory and flexible enough to account for changing conditions. Reciprocally, a prevailing legal framework is required that is enabling of collaborative and adaptive catchment management processes that meet these conditions, while not overly prescriptive and regulatory. The law should concern, facilitate and legitimise catchment-based planning, the operational aspects of management at all levels within the catchment, the application of relevant national policy at catchment scale, and monitoring and assessment. It follows that in the terms of our vision for catchment management proposed above, three questions are prompted.

1   What institutional arrangements under the law can facilitate the working of a programme that crosses sectoral and geographical jurisdictions through coordinated partnership-working at a catchment scale?
2   How can delegated roles and responsibilities remain well-coordinated vertically and accountable across levels of authority?
3   How can processes of knowledge-sharing with, and effective participation by, wider stakeholders be best encouraged and facilitated to enhance and legitimise catchment assessments, planning and implementation?

As Chapter 10 identifies in relation to the formation of collaborative catchment management programmes, there are few transferable prescriptions that can be made in answer to the first two of these questions. No single legal model can be prescribed for all conditions and jurisdictions, while in any given jurisdiction the institutional evolution of a successful collaborative catchment management initiative inevitably tends to precede codification of the framework of legal relationships and obligations that could facilitate it. To repeat the 'chicken-and-egg' analogy introduced in Chapter 10, the experience demonstrated by the case studies in this book suggests that the 'egg' of cross-sectoral and cross-boundary partnership-working precedes the 'chicken', i.e. the institutional framework, which could nurture it.

The case-study catchment management programmes in this book are thus important in providing examples of functioning solutions to this dilemma. The diversity of their organisational and institutional forms is summarised in the next section. Five institutional solutions are identified that may have at least some transferability to support catchment management initiatives elsewhere. Such measures are not mutually exclusive and may be effectively employed in combination. Their use will also be enhanced by collective development of strategic catchment, or regional cross-catchment plans, that serve to coordinate the contributions and actions of partner organisations. The five approaches identified are as follows.

•   Use of a Memorandum of Agreement or Understanding (MOA/MOU) as an innovative, flexible and formative device in the absence of existing legal provision. The purpose is to sufficiently formalise and operationalise joint and partnership working by diverse agencies with little prior history of close collaboration and coordination of actions. The application of MOA/ MOUs is illustrated in particular in Chapters 3 and 4 for the Upper Susquehanna and New York City Watersheds.

- Inter-municipal (or potentially other jurisdiction) agreements (defined here as a form of MOA/MOU with legal standing conveyed by higher statute or precedent), and as illustrated in particular in Chapter 5 for the communities of the Hudson Valley.
- Coordination and oversight by local government (subject to the capacity of its employed, contracted or partnered technical providers), and as illustrated in particular in Chapters 4 and 7 for the cases of Delaware County, New York, and partnerships between municipal administrations and water suppliers in Denmark, Germany and the Netherlands.
- Primary direction and coordination (though not regulation) by a higher-level government agency through dialogue, funding and technical support, as illustrated in particular in Chapter 5 for the case of New York State Department of Environmental Conservation (NYSDEC) and its leadership of the Hudson River Estuary Program (HREP).
- A networked membership organisation supported by a professional secretariat as illustrated in Chapter 6 for the Healthy Waterways in Queensland, Australia.

Specification of the detail of such institutional arrangements for any given jurisdiction is challenging. However, it is clear that such approaches are necessary. For land and water resource management there are simply too many decisions over time and space to be made, variables to consider, and issues to monitor, for top-down regulations to be comprehensively effective. Making rules and regulations executable at the local level where land, water and other natural resources are managed requires recognition that making regulation work will be enhanced by, and even dependent on, voluntary compliance. This is exemplified by the case of diffuse water pollution. For example, it is impossible for Defra in London, or for its Environment Agency in England, to closely regulate farmers' practices with regard to the application of fertilisers or pesticides on a continuing basis. It is similarly impractical for the US Environmental Protection Agency (USEPA) to continuously enforce the behaviour of dairy farmers in the Upper Susquehanna so that manure/slurry storage is adequate and manure is not spread on frozen ground to await the spring thaw to transport it to receiving waters. Thus, there has to be local and voluntary acceptance of regulation, and of the responsibilities and duties which this requires. Rules and regulations promulgated by agencies alone (including conditions attached to economic incentives) are always likely to be an insufficient mechanism for managing land and water resources. Intermediaries or technical providers able to work effectively and in partnership with farm businesses are essential.

The challenge is compounded by the reality that management practices may have a voluntary rather than regulatory basis. In the US for example, remarkably few environmental management measures for agricultural land are mandated by regulation. Under US federal law, there are only three clearly mandated environmental requirements: the Clean Water Act requires permits for concentrated animal feeding operations (CAFOs); pesticide uses are regulated under the Federal Insecticide, Fungicide, and Rodenticide Act; and the protection of listed endangered species is required under the Endangered Species Act. Rather than regulation, federal agricultural legislation favours voluntary programmes to encourage conservation and management measures to protect the environment. These voluntary programmes encourage the adoption and maintenance of good agricultural practices through economic incentives, supported by educational and technical assistance at the local level.

Non-governmental units including landowners and farmers, citizens' groups, non-profit organisations and businesses can all assist decision-making and delivery, by sharing priorities and information, and by contributing to practical actions. Some local organisations can be particularly effective as coordinators and intermediaries where they have built trusting relationships with other local stakeholders through repeated interactions, and they can provide an essential link between the public and government in representing citizens' concerns and priorities. However, although there may be an overall catchment framework for planning, and catchment-based partnerships for action, in practice diverse stakeholders are likely to mainly operate in different fora. For example, farmers are a distinct constituency with very different concerns to a land developer or a town planner. Intermediaries become essential to bridge such diversity in the absence of collective fora or functional channels of communication, and their role needs to be recognised within the institutional frameworks considered above.

In relation to a facilitating legal framework for collaborative and integrated catchment management, the third question posed above concerned how best to encourage and facilitate knowledge-sharing and effective participation by all stakeholders. Here, both international and national legal provisions are more clearly identifiable, although, at all levels, environmental law has increasingly emphasised procedural as opposed to substantive rules. Public participation and the engagement of non-governmental units and stakeholders, as described throughout this book, are a primary example.

A principal justification for increased participation is that it is a right in the sense that one is *entitled* to participate. This right is implied at the international level in the preamble of the UNECE Convention on Access to Information, Public Participation in Decision-making and Access to Justice in Environmental Matters (Aarhus Convention, 1998, p. 1). This states that

> in the field of the environment, improved access to information and public participation in decision-making enhance the quality and the implementation of decisions, contribute to public awareness of environmental issues, give the public the opportunity to express its concerns and enable public authorities to take due account of such concerns.

Article 1 of the Aarhus Convention (1998, p. 2) states as its objective:

In order to contribute to the protection of the right of every person of present and future generations to live in an environment adequate to his or her health and well-being, each Party shall guarantee the rights of access to information, public participation in decision-making, and access to justice in environmental matters in accordance with the provisions of this Convention.

Such emphasis on procedural rules illustrates how environmental legislation aims to provide a facilitating framework under which public participation is enabled. Similarly, Article Three of the EU WFD mandates cooperation within river basins at international and national levels in the EU (WFD, 2000); where adjacent countries share water resources, they must work together to institute sustainable management programmes. Also, in accord with the Aarhus Convention, Article Fourteen of the WFD requires EU member states to encourage participation by stakeholders in the implementation of the directive. Specifically in river basin management planning, the WFD requires member states to hold comprehensive consultations with the public and stakeholders to identify first the problems, and then the solutions. The citizens have a key role to play in implementing the WFD. The WFD assumes that public support and involvement is a precondition for the protection of water resources. Without public support in identifying both the problems, and the most appropriate measures to solve them including their costs, management measures will not succeed. Therefore, the WFD requires all interested and involved parties to be informed and consulted as river basin management plans are established.

At a national level, the WFD mandates EU member states to set up administrative structures to coordinate stakeholders whose interests are implicated by water resource management decisions. In its mandate for river basin authorities to prepare River Basin Management Plans (RBMP), Article Fourteen states that all interested parties must be involved in the discussion and plan creation. It states that public consultation is an essential element of the process and that RBMPs must be updated with the results of periodic public consultation. In England and Wales, for example, the WFD is made effective through 'transposing' legislation. For this purpose, the main transposing legislation is the Water Environment (Water Framework Directive; England and Wales) Regulations 2003 (Howarth, 2009). Under this legislation, for the purposes of the river basin management planning required by the WFD, the Environment Agency in England and Wales must provide opportunities for the general public, and also persons representing a comprehensive array of organisations identified in the legislation, to participate in discussion and in exchanging information or views, and to make representations regarding the draft plan for a river basin. Other EU member states have similarly assimilated WFD provisions into their national legislative frameworks.

Such legislative provision can provide 'top-down legitimacy' for a catchment-based collaborative approach. For example, since 2013 the

Department for Environment, Food and Rural Affairs (Defra) has introduced its 'Catchment-Based Approach' (CaBA) in England. Working at the catchment level this approach establishes partnerships of key agencies that work with stakeholders to agree and deliver the strategic priorities for the catchment and to support the Environment Agency in developing an appropriate River Basin Management Plan as required under the WFD. Objectives of the CaBA in England are:

1   'to deliver positive and sustained outcomes for the water environment by promoting a better understanding of the environment at a local level';
2   'to encourage local collaboration and more transparent decision-making when both planning and delivering activities to improve the water environment' (Defra, 2013, p. 4).

Adoption of the CaBA is expected to 'promote the development of more appropriate River Basin Management Plans (which underpin the delivery of the objectives of the WFD)' but also 'provide a platform for engagement, discussion and decisions of much wider benefits including tackling diffuse agricultural and urban pollution, and widespread, historical alterations to the natural form of channels' (Defra, 2013, p. 4).

As identified in Chapter 10, to be successful and sustainable it is equally important that catchment planning and management has 'bottom-up legitimacy'. Such legitimacy may be gained by establishing trust, respect and value for the catchment-based planning process and its outcomes, and is correspondingly served by legal means at the local level. Putting into effect such legal instruments is illustrated by the US cases in Chapters 3 and 4, by the Australian case in Chapter 6, and by the European groundwater protection cases in Chapter 7. Provisions for administration by designated regulatory or management agencies and partnerships must be 'fit for purpose' for the aims of higher and local legislation, and capable of effective delivery. At the time of writing it remains to be seen whether and how catchment-level partnerships established by Defra's CaBA can establish sufficient standing and legitimacy to become sustainable and fully effective.

## How to get things done: examples of institutional and organisational arrangements

The case studies in this book have evolved a variety of institutional and organisational arrangements to conform to the principles, characteristics and requirements described above, yet nonetheless demonstrate some common characteristics that exemplify how this can be done. As informs our vision for collaborative catchment management stated above, these include delegation of coordination to, or bottom-up initiation of coordination by, a local authority, government agency or other organisation, and partnership working between multiple organisations, agencies and levels of government.

Our cases would suggest that the most successful programmes establish a functional 'web' or network of partnerships supported by sound and transparent governance arrangements, access to funding, and scientific monitoring and modelling systems. These networks take account of the needs of catchment communities while working with them to achieve sustainable catchment management practices and meet the objectives of national policy. Some cases in this book also demonstrate the importance of scaling collaborative agreements and partnerships to the bio-physical and socio-economic regional demands of the catchment or river basin through inter-locality cooperation and coordination.

Yet questions remain as to how resilient and sustainable these arrangements are in the long term. From the bottom up, leadership, capacity and local coordination between partners can weaken if finance, continuity of employment or commitment to shared goals is uncertain. From the top down, policy drivers, funding, and political priorities and support, can shift or waver. If either happens, then horizontal and vertical coordination of actions, stakeholder commitment, and voluntary compliance with regulations and best practice recommendations can all be threatened.

Chapter 10 identified the dilemma of how to establish sufficient standing and influence with both decision-makers and stakeholders for a catchment management programme, while not assigning it the role of statutory authority and regulator. A solution to this dilemma requires the statutory authority responsible to recognise that if it disregards local concerns and priorities, or 'cherry picks from a local plan', it will fail to achieve the goals of its legislation and policies. For example, in the New York City Watershed (Chapter 4), the NYDEP first thought that it could impose a 'command-and-control' regime of regulation to protect its water supply but was persuaded to adopt a more inclusive and democratic process in the city's own interests. In 1998, the Delaware County legislature, which accounts for about 50 per cent of the New York City Watershed by area, formally adopted the Delaware County Action Plan (DCAP) through a resolution that also created a new county Office of Watershed Affairs with the responsibility to coordinate the DCAP. The DCAP operates within the framework of state and federal laws in full partnership with the agencies representing those governmental levels, and thus has some guarantee of institutional standing and sustainability. However, the uniqueness of the programme continues to raise many technical issues regarding legal and administrative requirements, organisation, communications, authority and responsibility. The US federal framework of water law has traditionally not recognised local governments as partners in environmental management, and this lack of specific recognition is reflected in the watershed MOA. Nevertheless, the MOA is intended in spirit and letter to incorporate a partnership of all levels of government in pursuit of shared watershed objectives. Experience subsequent to the signing of the MOA has clarified and demonstrated the capacities and interests of the various federal, state, New York City and watershed-level agencies, and this experience is a basis for an integrated and continuing partnership which fittingly incorporates the responsibilities of watershed communities.

DCAP thus succeeds as a locally managed programme directly accountable to local elected officials. Secure protection of unfiltered drinking water supplies from an inhabited watershed requires that farmers, businesses, community leaders and residents willingly manage the sources of pollution they control. Accordingly, the management of non-point sources of water pollution in particular is a matter for local government. DCAP's primary agencies are county departments that work cooperatively with regional (watershed), state and federal partners through coordination provided by the Delaware County Department of Watershed Affairs. Taken together this complex 'web' of cooperative partnerships provides a leading example of decentralised, yet multi-level and inter-agency collaborative environmental management. The respected, trusted and locally well-accepted technical capacity of county-level staff is a key resource.

On their part, local partnerships and action groups must also recognise that failure to comply with regulation and to deliver national objectives leads to loss of both legitimacy and access to technical support and funding. There is a co-incidence of interests, but mutual recognition of this, and development of shared strategic goals and action plans, will take time and iteration. For example, it was after years of attempted negotiation and disagreement that the 1997 MOA established the NYC Watershed Regulations[1] and the broader watershed programme that now protects the integrity of the sources of water to the city. These regulations were approved by the partner signatories to the MOA, rather than by specification by a higher-level government agency. As considered further below, making such solutions work also necessitates trusted leadership capable of bridging sectoral and disciplinary interests, and influencing policies and funding decisions.

In comparison to the DCAP a different institutional arrangement is illustrated by the legal recognition of the Upper Susquehanna Coalition (USC; Chapter 3). The USC was originally created at the initiative of some leading managers of county-based Soil and Water Conservation Districts. Their original aim was to facilitate coordination across the Susquehanna basin simply on the basis of informal trust and goodwill. Qualities of technical competence, neutrality, broad-based trust and local acceptance, transparent working and a track record of successful winning and delivery of grant-funded projects enabled work to be done in spite of a lack of formalised status. Growing success prompted the USC in 2006 to become formally established as a Conservation District Coalition through a Memorandum of Understanding (MOU). The MOU is based on New York and Pennsylvania state laws that allow Soil and Water Conservation Districts to enter into multi-district agreements. Each county administration represented by the Soil and Water Conservation Districts in the coalition signed the MOU giving the USC legal standing at that local level.

Cooperation between partners at many levels has been critical to the HREP's success (Chapter 5). The HREP establishes its legitimacy through its status as a

New York State government agency (NYSDEC) coordinated programme, yet it is not a regulatory agency or regional authority. It is part funder and wholly a pragmatic delivery organisation that stimulates formulation of shared strategic goals delivered by facilitated and coordinated action by others. Thus, its Action Agenda is not a conventional agency plan, but rather a vision for the Hudson Valley shared by government and citizens. The framework that it provides legitimises and facilitates partnership working and co-funding arrangements. Coordinated investment in scientific investigation, monitoring and public education is establishing a powerfully effective combination of a citizenry and municipal administrations knowledgeable about the ecology and natural resources of the Hudson, and primed to support actions that further improve land and water management.

For this book, the Hudson Valley provides a challenge because catchment management occurs at a much larger geographical scale than the other cases. However, Chapter 5 also provides the answer that management of land and water resources must simultaneously (and with coordination and complementarity) occur both at the scale of each tributary stream and at the scale of the whole river basin. Traditions and practices of governance in New York State do not favour such a solution, and a major challenge for the HREP in promoting integrated watershed management in the Hudson Valley has been conflict between the need for regional-scale interventions and community-based preference for self-governance. In general, prior attempts at integrated catchment or river basin management at such a large scale have involved creation of unitary regional governance structures such as river basin authorities. However, a range of factors, including the onerous coordination requirements and costs involved, have made success difficult to achieve in many cases. Chapter 5 illustrates that in New York State, solutions can be found through inter-municipal agreements, which offer a flexible means to enable municipalities to protect and manage tributary catchments across jurisdictions otherwise threatened by disconnected land and water management decisions.

The three USA case studies in this book represent watershed programmes comprising professional staff as established under New York State and County laws. They thus have clear 'top-down' legitimacy under law. However, staff members have no regulatory role. Rather, the full-time staff manage an in-depth direct working relationship with watershed constituencies strictly on a 'willing-recipient, willing-provider' basis. Generally, staff members are highly competent, experienced and well-trained professional technical providers who work readily with stakeholders. Staff members provide extensive outreach, coordination with other local, state and federal agencies, and foster public–private partnerships. It is important to emphasise that each programme depends for sustainability on the 'bottom-up' legitimacy conveyed by the continuing goodwill, trust and support from the catchment constituencies they serve. Despite being formally accountable to elected county legislators, and ultimately to New York State, in their operation the programmes serve at the pleasure of stakeholders. In this sense, stakeholders ultimately have control. Sustained success in fostering publicly valued watershed management provides key justification for continued funding support to the programmes from local, state and federal sources.

The substantive, instrumental and normative benefits that can be gained from partner and stakeholder participation were identified in Chapter 11 and are asserted by the provisions of the Aarhus Convention quoted above. To be comprehensively effective a catchment management programme must be participatory, meaning that all interested stakeholders can be involved or represented. Provisions for genuine and effective participation create a public forum for information-sharing, discussion and deliberation, and facilitate partner contributions to implementation. For example, the Hudson River Estuary Action Agenda (Chapter 5) achieves an effective integration of the interests of numerous stakeholders including state and federal agencies, municipalities, non-profit organisations, academic and scientific institutions, businesses, trade organisations, landowners and volunteers through an ongoing dialogue with the professional staff and technical providers. A key aspect of all three cases in New York State presented in this book is dialogue with landowners, local decision-makers and the general public. This provides for feedback and adaptability while maintaining trust and acceptance at the local level.

Chapter 11 also identified the importance of adaptive and analytic-deliberative processes for catchment management. A catchment management programme must responsively balance competing interests and problems. Dynamic flexibility is necessary to ensure that resources are being used efficiently and in accordance with both national objectives and locally agreed priorities. A programme's organisational and institutional arrangements and its action plans must be able to adapt to changing circumstances, informed by changes in knowledge. The HREP (Chapter 5) provides an example of how flexibility can be achieved through institutionalised provision for periodic review of goals and action agendas. Adaptability is enhanced through representation of all stakeholders in management decision-making, information-sharing and effective and open communication between stakeholders. An open dialogue enables stakeholders to continually and informally evaluate actions, and can prompt more formalised evaluations as deemed necessary.

Sharing some similarities of scale and scope of ambition with the HREP, the Healthy Waterways (HW; Chapter 6) in Australia illustrates how regional collaboration across the whole water cycle can protect water bodies and a healthier environment in a context of rapid economic development. Establishing the capability, institutionalised form and legitimacy of the HW involved partnership-building over two decades. From scientific studies and an improving knowledge base evolved shared strategies, and an adaptive cycle of action plans delivered by multiple organisations. This was paralleled by building stakeholder engagement, partnerships and institutional structures, with the support of monitoring, research and public outreach. Key factors were development of a common vision, goals and trust between participants, engagement of stakeholders, a strong scientific knowledge base and communication tools, public education and accessing diverse funding sources.

As a not-for-profit, non-government, membership-based organisation the HW's governance structure equates to notions of horizontal, self-organising network governance (Jordan and Schout, 2006) rather than a vertically integrated, top-down, agency-focused form. It operates across existing norms, and legal and political boundaries, through multi-level collaboration between government and non-government actors. Its secretariat (now Healthy Waterways Ltd.) performs the role of network manager or 'network administrative organization' (McGuire, 2006, p. 36), drawing in resources from partner organisations and steering their actions to achieve specific goals. HW is sufficiently inclusive in scope and membership that a specific action required can in principle, and usually in practice, be solicited and gained from the government agency with the relevant mandate and budget line. The private sector, NGOs and community-based groups can then help to fill the gaps and do things that government agencies cannot do, or cannot do well. The common strategy of HW partners provides the mechanism to negotiate and articulate both the strategies and actions to be implemented over a five-year timeframe. This planning and commitment is underpinned by good scientific understanding, condition and trend analysis of ecosystem health, and use of future predictions (scenarios). The concept of an adaptive cycle has applied equally to both the actions taken to manage land and water resources, and the organisational and institutional arrangements that have evolved to deliver these actions.

With similarity to other cases in this book, HW derives legitimacy from both its procedures and its outputs. In terms of procedures, the key is 'good governance' as demonstrated by transparency, accountability and inclusiveness of all stakeholder interests. Membership by Brisbane City Council, local governments and other government agencies, all subject to electoral or other forms of political accountability, also conveys legitimacy. In terms of outputs, legitimacy is achieved by demonstrating effectiveness in the delivery of improved environmental management at a regional scale. The willingness of government and non-government partners to make their internal planning processes and budget allocations at least partly subservient to the common aims and plans of the HW network also conveys legitimacy. HW's sustainability thus depends on sustaining the commitment of its members and on wide public recognition of the value of its work.

The HW experience also illustrates particularly well the challenges in seeking vertical integration and coordination in governance and actions, alongside horizontal integration and coordination with catchment- or basin-level partners. The need for the HW network to accord with higher-level policy could become constraining for a non-governmental network driven by environmental values and local economic priorities. As noted in Chapter 6 vertical coordination with higher governmental aims imposes costs in terms of the time, information, planning and reporting requirements and professional capacity required, but it also provides opportunities through winning political support and funding based on recognition that it provides a cost–effective delivery mechanism for national and regional goals. 'Output legitimacy' for its members and stakeholders is also derived when HW demonstrates that it can influence as well as follow Commonwealth and state policy directives.

HW thus seems to thrive in spite of, or perhaps because of, a lack of statutory underpinning and legal authority. Sustainability, however, cannot be taken for granted. Not everyone 'gets it' and why it works. Many professionals and officials prefer working according to narrow specialist mandates and guarding their allocated resources for 'self-set' purposes. This can often be a matter of protecting self-interest or one's 'turf'. Partnerships depend in large part on leadership, trust, personal relationships and information resources that need to be continually refreshed or renewed. Legitimacy derived from 'outputs' similarly requires a continual renewal in terms of upward reporting and outward public communications. Sustaining an adaptive management cycle requires continuing investment in capacity and innovation.

The cases from Denmark, Germany and the Netherlands (Chapter 7) show that the protection of groundwater can also not be delivered by a single authority, and requires the voluntary compliance of land managers. This feature is reinforced when programme aims evolve beyond solely drinking water safety to incorporate other land management and economic goals. Given the centrality of drinking water safety as a driver, water suppliers and local municipalities may often be the natural and most capable organisations to lead and coordinate programmes, but success still depends on partnership working with other relevant agencies and inclusive stakeholder engagement. Legitimacy can be gained from national 'safe' water legislation, the mandate of the water supplier, and its partnership with elected local government, but will be reinforced by broad-based public acceptance and support for programme governance arrangements, actions and outputs. Sustainability and political resilience of funding can be enhanced when programme synergies producing multiple benefits provide opportunities to combine funding streams and promote horizontal coordination of relevant environmental and economic development action plans.

Both strengths and potential limitations of a community-initiated and led catchment planning and management project are well illustrated by the WWF RIPPLE project (Chapter 8) and Loweswater Care Project (LCP; Chapter 9). Led by dedicated volunteers both programmes achieved a high degree of local standing and legitimacy. This was reinforced by partnership working, not least with statutory agencies, and by the synergistic and effective outcomes which this achieved. In both examples, the role of a bridging agent or neutral intermediary who invested significant time and energies in developing trust with and between project participants was important. Generating an atmosphere and an acceptance of collective responsibility was a key aim. In turn, this led to ways of working which solved problems at lower total cost, and with notably lower transaction costs.

Significant time and resources had to be committed to engage stakeholders and obtain their inclusive participation. The lesson is that catchment management coordinators and technical providers may need to hold a series of events such as meetings and workshops to build interest, trust and participation over time. Some individuals may need to be contacted directly and engaged in person before attending a public meeting. There is no single recipe for success but these cases demonstrate the value of looking for 'triggers' or 'hooks' that stimulate interest and motivation on the part of individuals or groups. Thus, there can be value in exhibiting and celebrating the social, heritage and cultural significance of a river or lake alongside its environmental and economic values. In the WWF RIPPLE project, communities within the catchment would initially tend to 'resonate' more with their local tributary or section of the river rather than the Ballinderry catchment as a whole. Sense of place can thus be very important, and attempts to develop a shared and strategic vision may need to build up from very local issues to a catchment- or river-basin scale.

Both cases show that participatory visioning, problem-framing and self-determination on the part of volunteer groups can lead to broader and more inclusive stakeholder engagement and a strong sense of project ownership. It should not be expected that this will necessarily be without conflict or dispute. The LCP in particular identified tensions in attempting to create a 'common vision' at the expense of allowing disagreement and heterogeneity to thrive in a group of stakeholders with different interests and perspectives. There was thus a need to balance the reassurance and sense of purpose that a community-based project can derive from having a 'common vision', with the continuing dynamic of allowing dissent and disagreement to be heard. There needs to be acknowledgement that thinking differently, or 'outside of the box', continues to be held as important and valued. Catchment management needs a forum in which controversial issues can be aired, opened up, critically examined and worked upon collectively on a basis of trust and partnership.

Lack of formalised status did not hamper the formative stages of the WWF RIPPLE project or the LCP, and again may indeed have been an advantage. Both could demonstrate 'good governance' and gain legitimacy at least locally and with partner agencies from the way in which they worked. However, both also ultimately faced the challenges of how to achieve continuity in operation and how to resource further actions once their initial project-based funding concluded. Both must seek to maintain and strengthen legitimacy based on their achievements, while gaining support and partnership from government and non-government organisations working at higher institutional levels and/or wider geographic scales.

To become welcomed as partners and delivery agents by such higher-level and wider-scale organisations is not a foregone conclusion. Another tension identified by the LCP in particular is that while individual representatives of other organisations could become enthusiastic about the LCP and its way of working, they could also remain unsure about how best to work together. Opening up knowledge-making and planning to stakeholder contributions and to their critical scrutiny can build 'buy-in' and support from a wide range of participants, but also create uncertainty for agencies that work to established mandates, patterns and procedures. This is not just about a threat of loss of control for an organisation, but also about creation of confusion, uncertainty and loss of confidence in the role or *modus operandi* of the organisation. Such uncertainty could, for example, equally affect an environmental NGO geographically structured by administrative areas such as counties rather than watersheds, or a government agency structured for top-down and programmed delivery of regulation, farm advice and grants for environmental improvements.

## Technical capacity and the role of technical providers

The examples of catchment management and groundwater protection in this book demonstrate the importance of the generation and application of scientific understanding developed through academic and other professional resources, and the availability of a professional cadre of technical providers able to work with and support stakeholders including farmers, other landowners, land developers, business interests and local governments. Although the variability in catchment circumstances makes it difficult to generalise about what technical capacities are desirable, the following capacities are common to the cases presented in the book. First, there should be a clear understanding of the applicable planning, environmental and land use laws. Second, commonalities in the technical capacities also evident in the cases are: an understanding of environmental conservation, land management, farm practices, highway and drainage or stormwater management, and river corridor management.

The Upper Susquehanna Coalition (USC; Chapter 3) provides a prime illustration of the importance of such technical capacities. Its strong reputation is based on technical competence and the effectiveness of diverse collaborative partnerships within and between different levels of government and non-governmental organisations. USC displays strong technical capacity at a local level and actions based on best-available scientific knowledge gained from its partnerships with the scientific and research communities. Local offices, locally relevant expertise and extensive efforts to inform and involve stakeholders and the wider public build broad-based trust in the USC, which is reinforced by transparency in governance and management, including full accountability for financial expenditure and performance.

There is a ubiquitous call in the relevant literature for more effective integration of science and decision-making in environmental management. The aim of such strategies is to enable decision-makers to access, value, consider and use expert opinion that they would not otherwise. Roles for intermediaries include convening, communicating, mediating and translating the best available scientific understanding for the fostering of participation, stakeholder interaction and the co-production of knowledge. It is desirable to bring together and integrate expert opinion from different sources and disciplines, and to structure, interpret and communicate scientific and other knowledge; all with recognition that knowledge itself can be a source of conflict in the policy realm. Different problems require different approaches, but it is the most difficult and controversial problems within catchment management that will require the most intensive and costly knowledge-brokering strategies. Indeed the role of intermediaries cannot be considered independently of the design of any participatory, analytic–deliberative and adaptive process and its anticipated outcomes. Their role and influence is integral in the necessary processes that build communication, trust, social capital and capacity.

The managers and scientific coordinators in what is now Healthy Waterways Ltd. (Chapter 6) provide a prime example of a team capable of operating at multiple levels with many partners to provide the intermediary roles essential in facilitating and coordinating a regional-scale collaborative catchment management programme. The HW secretariat has been the central intermediary of the HW network, but has also relied upon key collaborators in local universities, state agencies, local government and community groups. There has been explicit recognition of the need to ensure working practices that facilitate the development of personal relationships and long-lasting trust between parties. The HREP (Chapter 5) similarly depends on a number of organisations that play intermediary or broker roles. Apart from the NYSDEC itself, foremost among these in relation to tributary restoration (as well as other Action Agenda items) has been the Hudson River Watershed Alliance (HRWA), notable for its ability to facilitate dialogue and cooperation as a neutral and 'honest broker' without an adopted position on site-specific issues.

Chapters 3, 4, 5, 6 and 7 all strongly emphasised the importance of technical providers capable of delivery of technical support in catchment management and groundwater protection programmes. Chapters 3 and 4 highlight in particular the invaluable resource provided by the county-level Soil and Water Conservation Districts (SWCDs) in the USA. Generally trusted and accepted by land managers on the basis of their independent advisory rather than regulatory stance, they provide a local capability for the assessment of catchment condition and threats, for extension of best practice recommendations, and as a key intermediary with the rural community. As effective intermediaries they can, for example, facilitate farmers in interpreting and complying with regulation, and in accessing technical support and financial incentives made available under municipal, state or federal programmes. According to Helms (1992, p. 300) the districts 'accelerated acceptance of soil conservation in the United States by making landholders feel a part of the movement'. That 'movement' was not led by government alone, but also by farmers who influenced friends and neighbours to accept the values of conservation farming. Federal farm legislation now specifies conservation requirements for farmers who receive crop support payments and other assistance, but resource management problems cannot be solved by this instrument alone and local involvement and engagement remains essential (Helms, 1992). The ethos, traditions and *modus operandi* of SWCDs thus find themselves well-matched with the intent and practice of decentralised, collaborative and adaptive approaches to watershed management in the USA.

Helms (1992) also notes that the effectiveness of the 'districts' stemmed in large part from SWCD employees being drawn from the same backgrounds as farmers with similar values (most came from farm families and had qualifications in agriculture and environmental management from a local college or the state university). This was a decided advantage in seeking to encourage change in farming methods and adoption of best management practices. 'The fact that the "District" is operated by local people empowers them' (Helms, 1992, p. 301), enabling them to assert themselves as co-decision-makers in an equal relationship with land managers, and to accomplish more than could be achieved in any 'top-down' or paternalistic relationship.

These observations are echoed in the delivery of catchment improvement programmes in all of the cases in this book. The best farm advisers come from the same region and social background as the farmers of a catchment, speak with the same accent or dialect and share the same values and concerns about the local economy and environment. The best farm advice programmes also look at the farm manager and the farm business as a whole and do not focus solely on water protection issues. Their objective is to help ensure a sustainable and profitable business that meets the objective of the farm family, as well as catchment, regional and national objectives for environmental protection and stewardship.

## Leadership

A number of the cases described in this book identify the importance of leadership. This is particularly so during the early stages of the development of a catchment management programme, when formalisation of status may be lacking; similarly, legitimacy gained through demonstration of good governance procedures and delivery of outputs. Charismatic and energetic leadership can bring stakeholders together and facilitate partnership formation. Effective leadership can also work across disciplinary and administrative boundaries, often 'working both people and politics' to win funding and build information-sharing and collaboration that had not hitherto existed or proved functional.

Leadership is a critical ingredient in engaging stakeholders and in steering them through disputes and negotiations. Leadership can set and maintain procedures, build trust and facilitate dialogue. It can provide stimulus and motivation to seek to keep a process moving forward in spite of difficulties. Three key components of collaborative leadership are to:

1   manage the collaborative process;
2   maintain technical and scientific standards with credibility;
3   engage effectively with relevant higher authorities and delivery partners to ensure that the catchment management programme is empowered and that its decisions are implemented or have genuine influence.

Chapter 6 highlighted that these three components of leadership were important in the development of HW. Managerial leadership was provided by the network's governance structure and the HW secretariat. This facilitated collaboration between partners and wider public engagement. Scientific leadership was maintained by the status and authority of HW's Scientific Expert Panel and through the secretariat's emphasis on continual use of innovative modelling, monitoring and communication technologies. Lastly, political leadership and coordination was provided by the Lord Mayor of Brisbane and other leaders in local and state government.

Leaders must have, or recruit under their supervision, skills in stakeholder engagement and the facilitation of participatory processes, skills for monitoring, reporting and accountability, and skills as intermediaries as discussed above. Leadership is also required to help empower and represent weaker stakeholders in situations where incentives, power and resources are asymmetrically distributed or prior antagonisms are high (Ansell and Gash, 2008). Leaders will gain trust and respect and be granted legitimacy by collaborative partners based on qualities of integrity, competence and skills as a 'boundary-spanning' intermediary, although leadership status may also become formalised as governance structures evolve.

It was highlighted in Chapter 3 for the case of the USC that programmes need to plan in order to pass on and institutionalise leadership roles and skill sets. USC was created by and managed by exceptional leaders, not least the USC coordinator who in particular has played an innovative and effective role. As these leaders have approached retirement, or have assumed other responsibilities, USC has been deliberative in considering how to achieve effective transitions to new leadership. This is being achieved by nurturing emergent leaders who can assume leading responsibilities in a way that ensures the sustainable viability of the USC. This experience supports the view that leadership is not a matter to be left to fortune.

An individual or a team can provide leadership, and rotating leadership roles among partners can enhance 'buy-in', sharing of power or empowerment of weaker partners. A programme must be able to survive inevitable turnover in leadership, making provision for continuation of 'business as usual' during periods of transition and leadership change. Training and work experience should ensure that more junior staff members are prepared for future leadership roles. Formalisation of the standing and legal status of a programme, as discussed earlier in this chapter, may become necessary if operation has hitherto been dependent on the leadership and personal connections with partner organisations (particularly higher government) provided by one or two individuals.

A particularly important aspect of leadership is the ability to obtain funding for the catchment programme. Implementing a catchment plan generally requires substantial funding. Steps to implement a typical catchment plan will necessitate work and actions that cost money. New management practices on farms, better highway drainage schemes, establishment of open spaces, improvements in floodplain and riparian corridor management, or other implementation of components of the catchment plan, all incur costs. For example, under its leadership, the USC has an outstanding track record in annually obtaining substantial federal and state funding and grants for the technical services it provides, and for its multiple projects with farmers and other catchment partners. This substantial funding stream allows the coalition to 'get things done' including farm nutrient management, stream restoration, riparian buffer management, ditch management, highway and stormwater drainage management, and wetland restoration.

Finally, HREP demonstrates the multi-faceted nature of leadership in catchment management. The type of leadership is a function of the catchment problem being addressed. For example, Soil and Water Districts have a lead role with respect to the farming and rural landowning community in the Hudson River Valley. However, in the Hudson River catchment, the key constituency is land developers. Leadership with respect to this constituency is most likely to be provided by planning professionals.

## Outcomes

The concept of output or outcome legitimacy has been introduced above. How do we know that a catchment management programme works? What measures of effectiveness and success will justify its resourcing and continuation, and convey legitimacy? One can measure outcomes in terms of management policies and practices that are effectively adopted and implemented. A more fundamental measure of outcomes is the actual quantity, quality and character of the water resources of the catchment. As noted in Chapter 2 these challenging questions are difficult to resolve, and frameworks and empirical evidence for evaluation of the effectiveness of catchment management processes and their outcomes are typically poorly developed.

The challenge is particularly the consequence of the lag that typically occurs between improvements in management of potential causes of water contamination, and the consequent improvement in water quality. Confounding effects, which can include climatological variation, can wholly obscure improvements in water quality. Better water quality in groundwater or streams following improved management of diffuse sources of pollution may not become apparent for years. Likewise, measures to ameliorate the risks of flooding may have delayed or even ambiguous beneficial outcomes. Lags in water quality improvements are further confounded by the technical difficulties in attributing any water quality improvements to their specific causes.

Given the almost inevitable lags in measurable outcomes in a catchment's water bodies, and the indeterminate nature of their causes, the fall-back measure is to record the adoption of the management measures themselves. Simple examples are: the number of farms in the catchment that have adopted a nutrient management plan, measurable reductions in the applications of agricultural chemicals on farms, new manure storage and treatment systems, other structural measures such as improved swales, ditches and catchment basins, new or restored wetlands, etc. These are all tangible steps that can be assessed and recorded as measures of progress in implementing a catchment management plan.

The success of the USC (Chapter 3) can be demonstrated in terms of both process and outcome indicators. USC's programmes have gained a strong reputation for their actions to reduce flooding, water quality degradation and increase habitat and wildlife diversity. Basin-scale evidence for this is difficult to ascertain directly given the scale of the Susquehanna basin and the relatively high baseline water quality in the USC's region of operation, but the programme can point to many location-specific examples of demonstrable improvement. However, the Upper Susquehanna River is generally of excellent quality. Indeed, these headwaters have the highest quality in the entire Susquehanna River system within the Chesapeake Bay Watershed. This may be viewed as a tangible demonstration of successful outcomes. The established reputation and governance of the USC are also valid process indicators, given its acceptance by federal and state partners as the principal watershed-wide delivery partner for the Upper Susquehanna.

Chapter 4 cites the lifting of the 'phosphorus restricted' designation for the Cannonsville Reservoir basin as an environmental standards-based outcome indicator of the contributions made by the DCAP watershed programme. Over the years since the DCAP was established, the quality of water in the Cannonsville Reservoir has remarkably improved. As for the USC, many local examples of valued processes, actions and outcomes can also be identified. The New York City watershed is nationally and internationally cited as a model of upstream watershed protection financed by downstream water consumers and yet it remains controversial within the state. The watershed communities seek to use all indicators available to demonstrate that watershed protection by the residents of an inhabited and working landscape is cost-effective and preferable to either returning water supply catchments to 'wilderness' or construction of water treatment facilities to protect city supplies.

The outcomes of the HREP (Chapter 5) are numerous, diverse and spread across a wide geographical scale. In terms of process, developments in the capacity of state agencies, local governments and local communities as stewards of natural resources have been significant. In terms of environmental conservation, programme actions have enhanced many aspects of the natural environment of the Hudson Watershed, while contributing a stimulus to outdoor recreation and tourism and the regional economy. The Hudson River itself has improved over recent years as demonstrated by reinvigorated fish stocks such as sturgeon. A key to this success is the effectiveness of HREP's support for leaders appropriate for the outcomes sought. For example, land development and urban sprawl in the lower Hudson Valley have put strains on water supply capacities. Land development has also greatly augmented the risks of costly flooding in communities throughout the region. Controlling the risks of flooding is now a high priority. Accordingly, land developers have been successfully encouraged to adopt green development and construction methods.

Readily adopted measures including cluster and mixed land use development, the protection of open space, creation of riparian buffers, improved drainage and enhanced groundwater recharge, such as rain gardens and artificial wetlands, are all having an accumulative beneficial outcome. Planners and Soil and Water Conservation District professionals at the county and town level have been especially effective in promoting these measures.

Similar claims can be made for the region of South East Queensland by the HW (Chapter 6). Based on its comprehensive approach to monitoring and reporting it can claim success in terms of outcomes for specified pollutants, and in terms of process for levels of stakeholder engagement, changes in attitudes and partnership working. Challenges faced are, however, far from overcome as the programme strives to continue to restore and enhance inland waterway and coastal ecosystem health in a region undergoing rapid socio-economic development.

In terms of outcomes, the three groundwater protection programmes described in Chapter 7 are notable for their ability to demonstrate success in protecting safe drinking water supplies in regions of intensive agricultural production. In addition, an integrated approach to spatial planning has in each case been able to deliver a range of further environmental and socio-economic benefits. In terms of process, partnership working and well institutionalised governance arrangements appear capable of sustaining success and further building on achievements. Yet nothing is guaranteed. A tendency towards greater centralisation of governance and planning in Denmark and Germany and poorly 'integrated' renewable energy policies in Germany threaten past achievement or its future replication.

Chapters 8 and 9 mainly demonstrate success in terms of process indicators relating to stakeholder engagement and community-based planning. However, both engagement and planning have already stimulated some actions by farmers and other catchment residents and businesses. Some provisional indicators of improving water quality have been recorded but as yet cannot be attributed solely to these programmes.

Finally, it is important that the implementation of any catchment plans be evaluated in terms of determining not only the effectiveness of the implementation measures but also in terms of providing feedback that can guide mid-course corrections and adaption as experience and further knowledge are acquired.

## Conclusions: key components for sustainable catchment management

In this chapter, principles and commonalities for effective and ongoing collaborative and adaptive catchment management have been identified. These address as far as possible the legal framework that fosters legitimacy and accountability of catchment programmes, and facilitates partnership working. Ideally, that legal framework should encompass all levels of government, and corresponding law, including especially the local level. It should be facilitating rather than regulatory. For example, catchment management programmes in EU member states such as the UK that are inclusive of the stakeholders and provide for public participation are fostered at the international level through the Aarhus Convention, by the WFD, and potentially through transposition of the directive at the national level.

Successful catchment management has been shown to involve multi-level governance and the integration of both the 'top down' and the 'bottom up'. It requires vertical coordination of roles, responsibilities, actions, information flows and resources across levels of governance. It also requires decision-making and implementation at the catchment scale by multiple organisations and stakeholders that coordinate their actions horizontally in a coherent action programme. The partnership arrangements that make these processes work need to achieve sufficient standing and legitimacy to have both local acceptance and be well-assimilated in existing multi-level governance structures. Leadership and skilled intermediaries are needed to develop and sustain programmes, to assist in obtaining the necessary funding to meet the costs of implementing the catchment plans, and to facilitate coordination of actions across administrative boundaries and at larger geographic scales as necessary to account for physical and socio-economic interdependencies up to regional and whole-river-basin scale. Ways to measure and evaluate the outcomes of implementation need also to be applied to ascertain success and to inform an adaptive management cycle.

The realities at a catchment scale are complex and messy. For example, farmers differ in their interests and needs from other landowners. Land developers are often distinct in their priorities from environmentalists. Local governments are a distinct but primary stakeholder that should seek to represent all interests. Effective catchment management identifies, understands, and accommodates these multiple stakeholder interests by incorporating their disparate concerns and needs to ensure sound actions and sustainable outcomes.

## Note

1   An extensive list of rules addressing activities and contaminants threatening the water supply such as fertilisers, pesticides, stormwater pollution, wastewater treatment plants, radioactive materials and petroleum products (Chapter 4).

## References

Aarhus Convention (1998) *Convention on Access to Information, Public Participation in Decision-Making and Access to Justice in Environmental Matters*, UNECE, www.unece.org/env/pp/treatytext.html, accessed 8 August 2014.

Ansell, A. and Gash, A. (2008) 'Collaborative governance in theory and practice', *Journal of Public Administration Research and Theory*, vol. 18, no. 4, pp. 543–571.

Defra (2013) *Catchment Based Approach: Improving the Quality of our Water Environment. A Policy Framework to Encourage the Wider Adoption of an Integrated Catchment Based Approach to Improving the Quality of our Water Environment*, May 2013, Department of Environment,

Food and Rural Affairs, www.gov.uk/government/publications/catchment-based-approach-improving-the-quality-of-our-water-environment, accessed 8 August 2014.

Helms, D. (1992) 'Getting to the roots', in *People Protecting Their Land: Proceedings Volume 1*, 7th ISCO Conference, Sydney, International Soil Conservation Organization, Sydney, Australia, pp. 299–301.

Howarth, W. (2009) 'Aspirations and realities under the Water Framework Directive: Proceduralisation, participation and practicalities', *Journal of Environmental Law*, vol. 21, no. 3, pp. 391–417.

Jordan, A. and Schout, A. (2006) *The Coordination of the European Union: Exploring the Capacities of Networked Governance*, Oxford University Press, Oxford.

McGuire, M. (2006) 'Collaborative public management: Assessing what we know and how we know it', *Public Administration Review*, December 2006, pp. 33–43.

WFD (Water Framework Directive) (2000) *Directive 2000/60/EC of the European Parliament and of the Council of 23 October 2000, Establishing a Framework for Community Action in the Field of Water Policy*, European Commission, www.ec.europa.eu/environment/water/water-framework/index_en.html, accessed 8 August 2014.

# 13 Conclusions and future challenges

*Laurence Smith, Keith Porter, David Benson and Kevin Hiscock*

## Introduction

A healthy river, and in turn a healthy estuary and near-shore coastal zone, requires a healthy catchment in which uplands, riparian corridors, floodplains and wetlands are managed to steward and protect water resources. This requires management of land in ways that minimise movement of soil and pollutants to water bodies, minimisation of impervious surfaces, and sensitive design of dams, barriers and other modifications to natural stream morphology that cannot be avoided. It also requires efficient and sustainable management of agricultural production systems, including livestock, to minimise both import and export of nutrients to and from the system. Achievement of a healthy catchment will advance other natural resource conservation goals including protection of habitat and conservation of biodiversity. It will make both human and natural systems more resilient to the threats from climate change, and it can contribute to human health, wellbeing and economic activities based on recreation and countryside access.

In this book, the terms 'catchment' and 'watershed' have been used to refer to the sub-basins of tributary streams or to a whole river basin itself, as defined by the watersheds that divide drainage areas. Over-abstraction of water, flood risk and degradation of water quality are common and interdependent challenges in most inhabited catchments. Water moves, or is lost or stored, in ways that depend on land and urban infrastructure and how these are managed. This book's focus has been how best to protect and conserve water within landscapes which achieve the economic and social goals of the communities affected, in accord with higher-level policy and regulation. Of particular concern has been the deterioration in water quality caused by diffuse sources of pollution.

Three propositions have framed the assessments made in this book. First, that there are examples of catchment management programmes in a range of countries and situations that are showing evidence of success in protecting water resources at their source despite the complexities and difficulties of this challenge. Second, that despite the heterogeneity of catchment conditions, successful cases share commonalities in the processes of technical assessment and planning, actions and governance arrangements that can be identified through comparative analysis. Finally, that once identified and adapted to local conditions these commonalities can be codified as guidance for the development of catchment management programmes elsewhere.

This chapter presents guidance synthesised from the lessons of experience from the catchment management programmes presented in Chapters 3 to 9, and summary discussions in Chapters 10 to 12. The diversity of situations presented in this book demonstrates that this synthesis should not be overly prescriptive but presented in the form of guiding principles and commonalities of scope, scale, processes and governance arrangements.

## Commonalities for catchment management

Since there are no stopping rules or end points in solving complex environmental problems, decision-making must be based on adaptive, self-reinforcing processes that maintain effectiveness, momentum and legitimacy. The defining characteristics of such processes are rarely subjected to detailed analysis and made explicit. The aim of this chapter is to guide formation of catchment-based, collaborative, integrated and adaptive management approaches through:

- effective use of science and communication tools well-integrated with provisions for managerial decision-making to guide policy, planning and implementation of measures;
- collaborative partnerships, stakeholder participation and wider public engagement that direct, enhance and support decision-making, and provide a basis for co-generation of knowledge, knowledge exchange and sharing of best practice;
- decision-making and implementation at the level which is most effective and accepted within catchments.

In functional terms a catchment management programme should be able to:

- engage with and understand all stakeholder concerns;
- identify, scope and build working partnerships;
- scope preliminary goals and boundaries for assessment and planning;
- investigate and describe the key bio-physical and socio-economic variables in a catchment;
- refine management goals, targets and options;
- seek and secure funding in addition to any core public financial provision;
- plan implementation using dedicated funding and the resources of partners effectively and efficiently;
- plan and implement communication strategies for partners, stakeholders and the wider catchment community;
- monitor, evaluate progress and refine plans and implementation.

To achieve this, the following commonalities of scope, process, governance, capacity and evaluation criteria can be identified.

### Commonalities of spatial scope and boundaries

The term 'catchment' refers to the sub-basins of tributaries or the whole river basin itself, as defined by the watersheds that divide drainage areas. In some countries 'watershed' also refers to this basin or catchment land area. It is logical to take a river basin or catchment as the spatial unit for analysis, planning and management of land and water resources. The river and its tributaries are common to all parts of the basin and to its interdependent water users. Thus management on this basis has obvious appeal, not least when water is scarce and all demands for its use within a basin cannot be met. Both water scarcity and diminished water quality impose a need to manage water between its source and its sink, while the interdependence of human water uses with each other and natural processes requires holistic and integrated catchment-based management. A sub-catchment or a river basin is the natural 'forum' for assessment of resources and of opportunities and constraints, and a natural unit for strategic planning and management. Within a catchment water users and land managers can perceive their dependence on a common resource and understand their interdependence. This has the potential to provide a basis for the resolution of competition over resources or other conflicts of interest. In terms of obstacles it is usual that watersheds and thus catchment boundaries do not match existing administrative and political boundaries. Technical capability, leadership and coordination of actions that can cut across such boundaries are required, and under current governance arrangements this is usually best achieved through partnership working.

### An adaptive management cycle

The complexity, temporal and spatial scales, dynamics and inevitable trade-offs of catchment management necessitate an adaptive management cycle, collaboration between agencies and levels of government and a 'twin-track' of deliberative partner and stakeholder engagement supported by targeted scientific research (Figure 13.1 and Box 13.1). These needs are exemplified by the challenge posed by diffuse water pollution. This is a difficult challenge for public policy and may require innovation in approaches for its solution. The problems to be addressed are complex and conflicts of interest are inevitable; conditions under which a twin-track approach of scientific research and deliberative stakeholder engagement can deliver improved outcomes. Processes of deliberation with partners and with wider stakeholders, adaptive management and social learning can work when they build shared knowledge and the capacity for trust and collective action. These processes need to be supported by a sound scientific base, adequate financing, legitimacy and a facilitating regulatory environment.

Adaptive management is a way to accommodate the uncertainty inevitable in environmental management, and is a logical, systematic process for improving management by learning from the outcomes of policies and practices that have been implemented. It should be understood here as an iterative process of constructing assessments of catchment threats and problems, identification of potential solutions and learning from monitoring and outcomes.

Social learning, understood as the capability to transform a problem situation through change in understanding and practice (Ison and Collins, 2008), is an integral property of the deliberative processes depicted in Figure 13.1. This incorporates co-creation of knowledge with stakeholders, change in stakeholder perceptions and values over time, strengthening of analytical capacity and of social capital and trust between partners, and change in behaviour by individuals and organisations (Smith and Porter, 2010).

Although it presents a logical strategy, the development of an adaptive approach can challenge any organisations that suffer from a fear of failure and of being held accountable. Adequate time, skills, resources and political support are also required to allow maturity of the cycle of action and reflection to develop. Existing agency planners and managers may require training, and administrative and political support, to implement adaptive management.

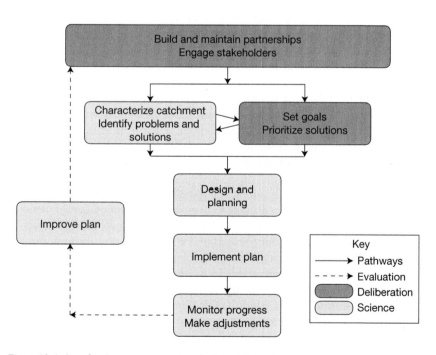

*Figure 13.1* An adaptive management cycle for catchment management

## Box 13.1 The adaptive management cycle in Figure 13.1

Figure 13.1 seeks to depict the following key features. It is intended as an indicative and stylised guide to action rather than a 'blueprint'. The first box in the diagram – 'Build partnerships and engage stakeholders' – is positioned to launch the process, and it is drawn broadly to suggest that it underpins and encompasses the whole cycle of activity. Its shading indicates that it is primarily a deliberative activity dependent on communication, discussion and negotiation. It should commence a process of visioning for the catchment, and the development of shared goals.

This first stage of building of partnerships and stakeholder engagement is critical but rarely receives adequate emphasis. 'Bringing together people, policies, priorities, and resources through a watershed approach blends science and regulatory responsibilities with social and economic considerations' (USEPA, 2008, p. 50). Successful development and implementation of a programme of catchment improvement depends on both partnership working, and on the acceptance, commitment and involvement of community members. It is critical to build relationships with key interested parties at the outset. Stakeholders include those who make and implement decisions, those who are affected by the decisions, and those who have the ability to assist or impede implementation; also those that can contribute resources and assistance, and those who work on similar programmes that can be part of an integrated or larger-scale effort. All of these, and not just those who volunteer, should be kept informed and given the opportunity to participate. As noted in this book, catchment management is usually too complex and expensive for one organisation alone. Partnerships can contribute new ideas and resources, increase public awareness and understanding of the problems, coordinate efforts and build public commitment. Initial identification and scoping of partnerships can avoid duplication of effort and economise on time and resource use.

In this first round of planning it is important to identify the interests of stakeholders. Their issues will shape local goals for catchment management and determine, at least in part, what types of data are needed. The goals initially identified may be quite broad but will be refined as characterisation and understanding of a catchment is developed. This relates to possible issues of conflict and power relations, and is all about collaborative framing of a shared understanding of the problem and the building of mutual commitment to its solution. The terms of reference, purposes and conduct of decision-making processes need to be agreed at the start of any consensus-building process (Sidaway, 2005). Shared understanding does not necessarily mean agreement but that stakeholders understand each other's interests sufficiently for effective dialogue about all interpretations of problems and for use of collective intelligence to address them (Conklin, 2006). The aim is to design a shared future through development of shared values that guide decisions, and all significant perceptions of a problem must be heard and acknowledged to ensure that the problem is fully understood and shared. Over time discussions can progress from current problems to scenarios that meet the collective values put forward, hopefully dissipating conflict arising from defence of current rights and benefits (Rogers, 2006). The vision created should be a future built on collective values which integrates stakeholder preferences, scientific knowledge, local experience and higher-level regulatory and policy directives, but in practice the process may need to move rapidly from 'visioning' to at least initial practical action if it is to sustain local stakeholder commitment (Smith and Porter, 2010).

It is difficult to generalise the specific actions needed for this first stage in Figure 13.1. For a new catchment programme initial community engagement and programme publicity using a variety of media and existing local social networks is important. This may then lead to a series of events such as public meetings and workshops between partner organisations. This was well-illustrated in this book by the activities of the WWF RIPPLE project (Chapter 8) and Loweswater Care Project (Chapter 9). In the three USA case studies (Chapters 3, 4 and 5) technical capacity existed in the form of professional staff employed by county- and state-level government. These technical providers develop programmes that depend for survival on the goodwill, trust and support from the catchment constituencies they serve. They also depend on partnership working with other agencies. These technical providers and their partners work closely with key stakeholders such as farmers and developers, and thus 'truth-test' and deliberate on actions as part of this process. This can be a more continuous and iterative process less requiring of events such as public meetings, although these may still be of value.

The next two boxes in Figure 13.1 highlight the essential and dualistic character of a 'twin-track' analytic-deliberative approach. Scientific evidence and modelling techniques (as reviewed in Chapter 11) are employed to describe the catchment and to identify problems and potential solutions, and this is paralleled by deliberation by partners and other stakeholders to set agreed goals and prioritise resource use. It is necessary to characterise a catchment to develop an understanding of the impacts of human activity that are present, identify the sources of these, and subsequently quantify pollutant loadings, over-abstraction of water, exacerbation of flood risk or other problems. This characterisation provides the basis for developing effective management strategies (USEPA, 2008). Catchment planning will not start from scratch and the aim will be to understand and build on existing information and actions, moving to prioritisation of needs and targeting of further data collection and analyses.

Characterisation informs the refinement of more precise management goals and associated targets that can guide development and implementation of a management strategy. Goal-setting and management planning will proceed iteratively, along with identification of indicators to be used to monitor progress towards meeting those goals. Potential management measures and management practices to achieve the goals will be identified. Options must be screened for those most promising and feasible, utilising both stakeholder knowledge of local conditions and increasing input from scientists, engineers, technicians and other professionals to ensure that the actions being considered are realistic and cost-effective in meeting land and water management objectives.

Appropriate problem-framing can integrate expert opinion from all interested parties, helping to ensure the relevance, credibility and legitimacy of research and recommendations. Use of models (Chapter 11) may become essential to the processes of deliberation to make complexity comprehensible and manageable. Conversely, deliberation is essential to the construction and use of models in setting priorities and goals, and in resolving trade-offs among outcomes. Stakeholders participating in an analytic-deliberative process can be best able to pose coherent, plausible and innovative solutions given their 'expert' understanding of local values and priorities, feasibility and cost-effectiveness. Again, such participation can occur in public meetings or workshops, or through close partnership working between technical providers and the stakeholders concerned.

Designing the implementation programme for the management options selected then generates other essential elements needed for an effective catchment programme. These include: an information/education campaign to further support public participation and build the management capacity of individuals and organisations related to adopted management measures; a work schedule for implementation of measures; interim milestones for monitoring progress; indicators of performance in relation to operational efficiency and the achievement of goals; budgets and resource mobilisation plans; specification of the legal authorities needed for implementation of measures; and monitoring activities and an evaluation framework (USEPA, 2008).

As work proceeds, part of the information/education programme should be efforts to publicise implementation of the catchment management strategy and its results. Regular communication is essential in building the credibility of and support for the programme, and transparency builds trust and confidence in the outcome. Regular communication also helps to strengthen accountability among partners by keeping them actively engaged. It might also stimulate more stakeholders to get involved, offering new ideas or resources. Sustaining political support and resourcing from state and non-state stakeholders will depend on the ability to demonstrate effectiveness and success. Communication tools such as an 'Ecosystem Report Card' (see the example in Chapter 6) let stakeholders know whether the status of water bodies are improving, and allow comparison of results between areas and over time. This can be an effective way to build awareness of catchment issues and the progress of improvements, and to sustain commitment.

Cost-effective monitoring is critical to an iterative adaptive management approach. Monitoring results need to be reported at the right time, place, quality and frequency to improve management. There is a need for decision points at which information can be reviewed and changes to the programme made. A conventional adaptive management cycle of planning and implementation, and monitoring and evaluation that feed back into planning is thus depicted in the final stages of Figure 13.1. However, note that the use of shading maintains emphasis on processes of both continued technical analysis and inclusive deliberation.

Consideration of Figure 13.1 prompts the question of who should lead or coordinate this cycle of activities. There is no single answer to this. The role is played by technical providers employed by local government in Chapters 3, 4 and 7 of this book; by a US state-level agency in Chapter 5; by the professional secretariat of a networked membership organisation in Chapter 6; and by community-based NGOs in Chapters 8 and 9.

### Governance arrangements

Governance arrangements for catchment management include the range of political, legal, economic, administrative and social systems that are in place to manage a catchment's land and water resources. They thus encompass the responsibilities and actions of different levels of government and government agencies, private businesses including farms and civil society organisations or community groups.

### Meaningful and sustained opportunities for public participation

The case studies in this book have demonstrated how provisions for genuine and effective participation can create a public forum for information-sharing, discussion and deliberation. This mechanism is essential to ensure that achievement of national environmental and public health criteria typically set by 'top-down' regulation or policy can be integrated with achievement of local economic and social goals, including sectoral goals such as those of catchment farming, fishing, tourism or other industries. Knowledgeable local stakeholders can also contribute to catchment assessments and programme design, and implementation will be enhanced by local knowledge, acceptance and sense of ownership. As noted above, and as illustrated in the case study chapters, there are a range of processes and mechanisms through which these benefits of participation can be gained.

*Collaborative partnerships within and between levels of government, sectoral and area responsibilities, the private sector and non-governmental organisations*

The case studies in this book show that the organisational structures and institutional arrangements of catchment management partnerships are diverse and there is no single model that can be proposed. However, the case studies do suggest that catchment management programmes should be built from existing organisations and partnerships, centred on those with current management responsibilities, and working within the framework of prevailing law. The building of partnerships must establish shared goals and recognise differentiated interests and responsibilities. A comprehensive and holistic approach to catchment management requires technical capability, leadership and capacity for coordination covering at least agriculture, water supply, wastewater and waste management, highway and other storm runoff, stream corridor restoration, and development and spatial planning. There is a need for genuine and proactive 'joined up thinking' between all the parties who have a significant role, at whatever level, in catchment management.

This central need can similarly be expressed in terms of the requirements for an integrated approach to catchment management. Such an approach requires partnership working to achieve:

- spatial integration of management of land, waterways and water bodies; including estuaries and near-shore coastal ecosystems;
- integration of management of water quantity and quality;
- integration of national, local and sectoral goals;
- integration of scientific research and assessment with the local knowledge and experience of catchment stakeholders, and of the resulting combined information with policy formulation and management decision-making;
- integration of technical expertise across scientific disciplines and sector specialisms;
- horizontal integration and coordination of actions at catchment scale by government, business and civil society.

Laws and other rules or codes of practice are thus needed that facilitate rather than prohibit partnership arrangements, coordinated actions and appropriate delegation of responsibility. These may include, for example, provision for regional strategies and action plans, backed up by inter-locality agreements to facilitate and coordinate required actions; and provision for knowledge acquisition, scientific peer review, knowledge-sharing and synthesis, capacity-building and public education at a catchment or even river-basin scale. It has been noted in Chapters 10 and 12 that a Memorandum of Agreement or Understanding (MOA/MOU) may provide a flexible legal instrument to support the establishment of partnership working to these ends.

It will frequently be the case that, at the most local level, important partners exist in the form of community groups comprised primarily of citizens and local businesses including farms. Such assemblies of stakeholders typically convene (periodically or regularly) to discuss, negotiate, plan or implement the management of streams, river channels or sub-catchments (including land-based measures designed to manage water quantity or quality) in their locality (Chapters 8 and 9 in this book provided examples). Common characteristics of such groups are:

- use of catchment boundaries (at various scales) as units for assessment and management;
- concern for a range of issues including stream flows, water quality, water use and demand management, aquatic and riparian habitat, landscape amenity and heritage, recreational access, livelihoods and rural economy;
- participation by multiple local and non-governmental interests in seeking to influence decisions by local government and government agency representatives;
- assessments and decision-making for the locality that draw upon bio-physical science as well as socio-economic information and local knowledge;
- an orientation towards collaborative planning and problem-solving based on the seeking of consensus and design for situation-specific actions;
- administration and action largely completed through volunteer contributions of labour and other resources.

Depending on the scale of a programme, a coordinator of collaborative catchment-based management must seek to draw on the strengths of such community-based groups while preserving their autonomy and self-determination.

*Legitimacy and institutionalisation of programme status*

In terms of appropriate delegation our introduction to this book (Chapter 1) and assessment of case studies (Chapters 3–9) suggest that local management of land and water at a catchment or sub-catchment scale should be recognised as the natural and default situation given its inherent logic and advantages. Chapter 10 further identified that an essential and core part of this logic stems from the characteristics of legal systems and the prevailing law.

Integrated land and water management involves local responsibilities and requires inclusive deliberation at the local level under the framework of existing multi-level government. Thus a catchment management programme needs to demonstrate that higher-level laws and regulations have been translated into locally acceptable duties, responsibilities and rights, supported by provisions for inter-locality coordination and cooperation where required.

Informal partnerships with effective leadership are often a starting point but growth in funds, capacity, authority and demands from higher levels for delivery will often necessitate steps to reinforce legitimacy and a formalised legal status, so as to sustain the standing of the programme in the eyes of its partners, stakeholders and higher authorities.

How to build and sustain such local partnership arrangements with legitimacy, sufficient standing and sustained funding is a core challenge for catchment management initiatives and national policy in this respect. It is, for example, the leading challenge facing Defra and the Environment Agency in their development of the 'Catchment-based Approach' in the UK. As we have argued in this book, government departments alone cannot achieve the water resource protection objectives set by EU or other national/federal directives or legislation. A broad societal response well-embedded at local level is needed, and it is the mobilisation of the understanding, commitment and resources of a catchment's own agencies, businesses and residents that provides the means to put policy into practice.

As discussed in Chapters 10 and 12 legal dilemmas remain in terms of how best to institutionalise or codify such delegated and partnership-based catchment management arrangements. This is related to questions of how to balance the independence, trust and role as facilitator of a coordinating catchment management organisation with the need for standing and legitimacy; and how to secure and sustain the commitment and participation of stakeholders in deliberative assessments, planning and joint action if there is no guarantee of acceptance, and thus in effect delegated decision-making power, from higher authorities. One answer is to adopt an adaptive approach and allow time for organisational and institutional arrangements to evolve based on the trust and support which their well-chosen actions can solicit and inculcate from both catchment stakeholders and higher authorities. Another answer, as proposed in Chapter 10, is to make initial use of flexible legal instruments such as MOU/MOAs.

### Transparency and accountability

Principles of good governance emphasise that decision-makers should be accountable for their responsibilities. Thus, all data, synthesised information and decision-making of a catchment management programme should be available to the public and open to scrutiny. Key actors must assume and be accountable for their delegated responsibilities and outcomes. Ultimately, achievement of such accountability through election of officials, or at least oversight by elected local government, may be the preferred options, but as illustrated by the cases in this book some diversity of organisational and institutional arrangements that achieve sufficient accountability is possible.

*Funding*

As illustrated by the case studies in this book successful catchment management programmes demonstrate the ability to access diverse funding sources including voluntary contributions and the private sector. However, many of the benefits of improved catchment management take the form of non-marketed or even at least partial 'public' goods. Thus public-sector funding remains a requirement for at least a proportion of catchment management measures, while continuity in institutional development and capacity-building can be expected to require core public funding, at least in the formative stages of a catchment management programme.

## Capacity components

A catchment management programme must demonstrate technical capability, effective and accountable leadership, functional coordination for catchment areas that do not correspond to administrative boundaries and broad-based public support. Thus programmes require a range of essential capacities to achieve their aims. Evaluation of the case studies in this book reveals the following to be the most essential.

*Leadership*

Chapter 12 emphasised the importance of managerial, technical/scientific and political leadership for the initiation, development and sustainability of a catchment management programme. There is a need to identify and nurture emergent leaders, no mean challenge, particularly if seeking to develop catchment management partnerships across catchments at national scale. There is also a need to recognise that leadership can be multi-faceted and provided by different individuals or organisations in relation to their specialist roles and capacities within a coordinated programme. As they evolve, programmes also need to plan in order to pass on and institutionalise leadership roles and skill sets.

*Mobilisation of locally accepted technical providers*

Trusted individuals, agencies or groups are needed for capacity-building and advisory work, not least with farming communities. Their essential functions include convening and mediating to foster trust, participation, collaboration and co-production of knowledge. The leading example given in this book is that of the county-level Soil and Water Conservation Districts (SWCDs) in the USA, whose mission is to develop and oversee effective soil and water conservation and agricultural non-point source water quality programmes with the involvement of citizens and agencies, through education, partnerships, planning and implementation. However, the capacities of the SWCDs are mirrored in those intermediaries described in the cases from Australia, Denmark, the Netherlands, Germany and Northern Ireland in this book.

*Capacity to conduct comprehensive condition and threat assessments, and strategic and action planning, based on sound science and best available knowledge*

Catchment management programmes must be able to make assessments of the condition of, and all threats to, water resources and prepare comprehensive and integrated plans. Adequate technical capacity is needed, not necessarily to complete all scientific assessment and planning 'in-house', but to be able to work effectively with other suppliers of technical expertise and scientific peer review. Ideally all partners will come to agree and refer to one integrated plan, or at least strategic planning guidance, for the catchment. Planning and implementation must be based on credible science, and thus there must be the capacity to commission external expertise and scientific peer review.

*Capacity for monitoring of performance and outcomes*

Monitoring and evaluation of the processes and outcomes of catchment management is essential to the learning and responsiveness inherent in an adaptive management cycle, and for determination of the effectiveness and efficiency of outcomes. Reporting on governance, achievements and outcomes is also inherent to sustaining stakeholder and partner engagement, and to demonstrating the benefits of collaborative and integrated catchment management to higher authorities and the wider public.

*Capacity for knowledge exchange*

Programme technical providers need to act as brokers to compile, synthesise and communicate information, enabling decision-makers to consider and use diverse data sources. Gaining the benefits of partner and stakeholder participation in terms of enhanced assessments, planning and implementation requires an accessible knowledge base, skilled intermediaries and high-quality communication and decision-support tools. Wider public education and awareness-raising activities about catchment water resources and the aquatic environment can be a facilitator of commitment, support and action, and can become a two-way process. Such communication can be targeted at children through support to school curricula, as well as parents and communities.

The role of knowledge management is an often under-emphasised aspect of both public participation and catchment management. The generation and management of knowledge are critical to the process. If protection of water quality ultimately depends on the actions of land users and other actors, scientific knowledge will need to be integrated and delivered to help them sustain their economic interests and protect the environment. Key individuals and community representatives may need to participate in scientifically and technologically complex decision-making, creating challenges in how best to inform stakeholders about environmental risks and facilitate an informed discourse about issues and priorities. Citizens are capable of participating in management of complex natural resource problems (Fischer 2000; Sabatier *et al.*, 2005), but central to this capability are an accessible knowledge base and high-quality informational tools and decision-support systems. These need to communicate technical information to lay participants and effectively integrate scientific measurements with the location-specific contextual detail that local people can provide.

### Aims and outcome criteria

It is important to specify the aims and outcome criteria that should guide catchment management initiatives and their subsequent evaluation in any given context. Such aims and outcome criteria should be adapted to context by local deliberation but in most cases will include the following as central objectives.

#### Delivery of long-term water quality improvements and sustainable management of water resources

The ultimate goals of catchment management are to sustain agreed and designated uses of land and water in a catchment with a functioning ecology and sustained provision of desired ecosystem services, including provision of safe drinking water supplies, water supply for farming and industry, food and fibre from agriculture and forestry, conserved biodiversity and managed flood risk. Specification of these goals must also take account of inter-generational needs and seek to guard the future against degradation arising from present uses of land and water.

#### A viable rural economy and communities

The environmental goals above must be sought while seeking to maintain the viability and economic growth of a catchment's economy. Typically dominated, at least in terms of land use, by farming, tourism and recreation, a core challenge for catchment management programmes is sustaining the socio-economic vitality of an inhabited watershed while delivering cost-effective water resource protection.

#### Cost-effectiveness and efficiency in the delivery of outcomes

To be able to demonstrate cost-effectiveness and efficiency in the achievement of desired outcomes requires that there be an agreed prioritisation of needs and targeting of resources based on catchment assessments and plans. In planning there must be flexibility in policies and their implementation so as to achieve well-adapted local solutions. The evidence produced by monitoring, evaluation and reporting should also demonstrate cost-effective and efficient delivery compared to alternative approaches.

### Assurance and acceptance of the burden of costs and distribution of benefits

To minimise conflicts and to seek to ensure the sustainability of catchment management arrangements and improvement actions, it is important that the allocation of catchment resources is based on all existing legitimate interests and values and can be accepted as fair and equitable by all parties concerned. Here 'equitable' refers to what is both valid under the prevailing law and considered just and reasonable (though not necessarily equal). Similarly, there must be an equitable and agreed allocation of financial and other costs to carry out catchment improvements and to sustain desired catchment management outcomes.

### Process criteria

As discussed in Chapter 12 it can be difficult to evaluate preferred outcome criteria for a catchment management programme, particularly in the early stages of its development. Thus a parallel focus on process criteria is also important. The following list provides a non-exclusive guide to some key process criteria for catchment management programmes and includes an ability to:

- build or strengthen partnerships;
- establish a shared vision and feasible goals;
- engage in public education and awareness-raising;
- implement inclusive participation/deliberative engagement;
- legitimately represent local concerns and priorities;
- act in accordance with existing law and policy;
- accept responsibility for implementation of higher-level policy directives and regulation;
- prepare a comprehensive and integrated plan;
- manage knowledge effectively, with transparency and accessibility;
- raise and receive funding;
- coordinate and implement delivery;
- implement an adaptive management cycle;
- monitor and report outcomes;
- contract scientific support and peer review;
- resolve conflicts;
- demonstrate adoption of best practice measures for water resource steward-ship and protection, farming and other sectors.

# In conclusion

In this book we have assembled propositions based on real-world examples that catchment management requires actions across sectors, scales and existing administrative boundaries. It thus requires multi-level collaboration involving relevant government agencies and other organisations. In fact, catchment management must incorporate a full spectrum of decision-making from the central policy-making level, through regional and local instruments of governance, to the decisions of land managers. A sustained and successful catchment management programme depends upon acceptance and co-ownership of responsibilities between the catchment partners concerned.

There must be clarity on the allocation of roles, responsibilities and authority both 'vertically' between levels of government and 'horizontally' at each level. To be integrated, comprehensive and sustainable, catchment management has to be built from partnerships and collaborations that recognise the different levels of law and decision-making authority that apply. This requires examination of the balance between national and local government, and also the parallel emergence of community-based and other non-governmental environmental improvement initiatives. Determination is needed of how to foster vertical and horizontal coordination, and an assimilation of duties and responsibilities in a partnership between national and local governments and civil society organisations that can offer a unified and integrated catchment programme. Key elements include the legislative and policy provisions needed to support such programmes, which need to facilitate effective and balanced complementarities in the use of regulation, voluntary initiatives and economic incentives in changing the behaviour of land users and other actors who impact upon water resources. Knowledge management is an important feature. Information should flow between levels, organisations, stakeholders and the wider public to facilitate shared understanding, minimise conflict, garner support and coordinate action.

These characteristics and requirements pose challenges. First, administrative and legislative reforms or modifications may be required if the procedures and functions required at local level cannot be readily and formally assimilated within existing higher-level legal and administrative structures. It may be necessary to translate supra-national and national frameworks of laws and regulations into locally acceptable duties, responsibilities and rights, supported by provisions for inter-locality coordination and cooperation where required. Creation of the 'bridging organisations' or intermediaries that facilitate collaboration across different locations and scales is also critical. It may also be necessary to invest in strengthening human and organisational capacity, accountability and technical capability at local level. The need is to create technically capable, effective and accountable local leadership and functional coordination for catchment areas that will not usually correspond to administrative boundaries.

For these challenges working 'with the grain' of existing governance structures will be more practical and feasible than radical reform. Efforts should seek to build from existing organisations and partnerships, strengthening and expanding these as necessary. Application of existing law is always preferable to the creation of new law (Negro and Porter, 2009). The successful implementation of collaborative approaches to catchment management can be constrained by inadequate resourcing and a lack of human capital, but also by a lack of social capital, the exercise of power, and weak legal and political embeddedness. In summary, the role for regional or central government is to set the essential legislative and regulatory framework and seek to foster more partnership-based, participatory and adaptive approaches to problem-solving and implementation, achieving this through local instruments, participation of stakeholders and capacity-building, supported by sound scientific understanding and an enabling policy environment.

This formulation of catchment management can be portrayed as a paradigm shift. The literature on stakeholder participation and collaborative approaches for environmental management is extensive, but it focuses on 'bottom-up' initiatives and generally fails to systematically address what entities at each level make decisions, that together provide the basis for genuinely integrated catchment management. In contrast, earlier attempts at river basin management tended to remain relatively 'top-down' and focused on creation of regional authorities with cross-sectoral and cross-boundary mandates that could rarely deliver effectively. Thirdly, the Integrated Water Resources Management (IWRM) paradigm[1] has been widely adopted internationally, but is both broader and looser in scope than the catchment-based, collaborative and adaptive approach considered in this book. Although IWRM has also been widely criticised as a normative paradigm that lacks practical prescriptions and empirical evidence of its benefits, it remains a framework under which a collaborative and adaptive catchment management approach can be nested.

This book documents the success to date of some leading catchment management examples. The resulting paradigm for catchment management that is investigated and summarised as guidance sets modest and well-focused aims and it identifies practical commonalities for scope, scale, processes and governance arrangements. We hope this book helps facilitate the development of improved water resource management wherever this is needed.

## Note

1   Conventionally defined as 'a process which promotes the coordinated development and management of water, land and related resources in order to maximize the resultant economic and social welfare in an equitable manner without compromising the sustainability of vital eco-systems' (GWP, 2000, p. 22). It is a process in large part based on application of the Dublin Principles (ICWE, 1992).

# References

Conklin, J. (2006) *Dialogue Mapping: Building Shared Understanding of Wicked Problems*, Wiley, Chichester.

Fischer, F. (2000) *Citizens, Experts, and the Environment*, Duke University Press, Durham.

GWP (2000) *Integrated Water Resources Management*, TAC Background Papers, Technical Advisory Committee, Global Water Partnership (GWP), Stockholm.

ICWE (1992) *The Dublin Statement on Water and Sustainable Development*, International Conference on Water and the Environment (ICWE), Dublin.

Ison, R. and Collins, K. (2008) *Public Policy that Does the Right Thing Rather than the Wrong Thing Righter. Analysing Collaborative and Deliberative Forms of Governance*, 14 November 2008, The Australian National University, Canberra.

Negro, S. and Porter, K. S. (2009) 'Water stress in New York State: The regional imperative?', *The Journal of Water Law*, vol. 20, pp. 5–16.

Rogers, K. H. (2006) 'The real river management challenge: Integrating scientists, stakeholders and service agencies', *River Research and Applications*, vol. 22, pp. 269–280.

Sabatier, P. A., Focht, W., Lubell, M., Trachtenberg, Z., Vedlitz, A. and Matlock, M. (eds) (2005) *Swimming Upstream: Collaborative Approaches to Watershed Management*, MIT, Cambridge, MA.

Sidaway, R. (2005) *Resolving Environmental Disputes: From Conflict to Consensus*, Earthscan, London.

Smith, L. E. D. and Porter, K. S. (2010) 'Management of catchments for the protection of water resources: Drawing on the New York City Watershed experience', *Regional Environmental Change*, vol. 10, no. 4, pp. 311–326.

USEPA (2008) *Handbook for Developing Watershed Plans to Restore and Protect Our Waters*, United States Environmental Protection Agency, Washington DC, EPA 841-B-08-002.

# Index

For Product Safety Concerns and Information please contact our EU
representative GPSR@taylorandfrancis.com Taylor & Francis Verlag GmbH,
Kaufingerstraße 24, 80331 München, Germany

Printed and bound by CPI Group (UK) Ltd, Croydon, CR0 4YY
08/05/2025
01864513-0001